I0056589

Textbook of Biotechnology

Jordan Goldberg

Larsen & Keller
www.larsen-keller.com

Textbook of Biotechnology
Jordan Goldberg
ISBN: 978-1-64172-612-2 (Hardback)

© 2022 Larsen & Keller

⊟ Larsen & Keller

Published by Larsen and Keller Education,
5 Penn Plaza,
19th Floor,
New York, NY 10001, USA

Cataloging-in-Publication Data

Textbook of biotechnology / Jordan Goldberg.
        p. cm.
Includes bibliographical references and index.
ISBN 978-1-64172-612-2
1. Biotechnology. 2. Genetic engineering. 3. Chemical engineering I. Goldberg, Jordan.
TP248.2 .T49 2022
660.6--dc23

This book contains information obtained from authentic and highly regarded sources. All chapters are published with permission under the Creative Commons Attribution Share Alike License or equivalent. A wide variety of references are listed. Permissions and sources are indicated; for detailed attributions, please refer to the permissions page. Reasonable efforts have been made to publish reliable data and information, but the authors, editors and publisher cannot assume any responsibility for the validity of all materials or the consequences of their use.

Trademark Notice: All trademarks used herein are the property of their respective owners. The use of any trademark in this text does not vest in the author or publisher any trademark ownership rights in such trademarks, nor does the use of such trademarks imply any affiliation with or endorsement of this book by such owners.

For more information regarding Larsen and Keller Education and its products, please visit the publisher's website www.larsen-keller.com

Table of Contents

Preface

Biotechnology is a field of biology that makes use of living systems and organisms to develop products. It is a broad field that includes principles from the fields of genomics, immunology and recombinant genetics. It is also used in the development of pharmaceutical therapies and diagnostic tests. Some of the major branches of biotechnology are bioinformatics, green biotechnology, violet biotechnology and yellow biotechology. Green biotechnology refers to the application of the principles of biotechnology to agricultural processes. The issues of philosophy, law and ethics related to biotechnology are dealt with under the sub-domain of violet biotechnology. The utilization of biotechnology for the purpose of food production is referred to as yellow biotechnology. Major sectors in which biotechnology is applied are health care, food production and agriculture. This book provides comprehensive insights into the field of biotechnology. Most of the topics introduced in this book cover new techniques and the applications of this field. It will provide comprehensive knowledge to the readers.

To facilitate a deeper understanding of the contents of this book a short introduction of every chapter is written below:

Chapter 1- The field of biology that makes use of living systems and organisms for developing and making products is known as biotechnology. The branches of biotechnology include medical, animal, agricultural, environmental and industrial biotechnology. This is an introductory chapter which will introduce briefly all the significant aspects of these branches of biotechnology.

Chapter 2- Any piece of DNA which has been created by combining at least two strands is known as recombinant DNA. They are formed using the methods of genetic recombination for the purpose of combining genetic material from different sources to create sequences which are not found in the genome. This chapter discusses in detail the processes and applications related to recombinant DNA technology such as DNA cloning and DNA replication.

Chapter 3- The direct manipulation of an organism's DNA to change its characteristics in a specific manner is known as genetic engineering. Medicine, industrial biotechnology and agriculture are the various fields where genetic engineering is applied. The diverse applications of genetic engineering in the current scenario have been thoroughly discussed in this chapter.

Chapter 4- The branch of agricultural science which makes use of scientific techniques and tools to modify living organisms such as plants, animals and microorganisms is called agricultural biotechnology. Genetic engineering, tissue culture and molecular markers are some of the techniques used under this domain. All the diverse applications of agricultural biotechnology for the purpose of plant breeding and gene transfer in plants have been carefully analyzed in this chapter.

Chapter 5- The branch of biotechnology which makes use of molecular biology techniques to genetically engineer animals is known as animal biotechnology. Transgenic animals are a category of genetically engineered animals whose genome has been modified by the transfer of a gene from another species. The diverse applications of animal biotechnology have been thoroughly discussed in this chapter.

Chapter 6- The branch of biotechnology which uses biological processes to tackle environmental problems such as the removal of pollution and biomass production is known as environmental biotechnology. Bioremediation is an example of environmental biotechnology where microorganisms are used to consume and break down environmental pollutants. This chapter discusses in detail the theories and methodologies related to environmental biotechnology.

Chapter 7- Nanobiotechnology is the intersection of nanotechnology and biology. It aims to apply nanotools to relevant medical/biological problems and refine these applications. It is also focused at generating cures and regenerating biological tissues. Science and technology have undergone rapid developments in the past decade that have resulted in the discovery of significant tools and techniques in the field of nanobiotechnology; which have been extensively detailed in this chapter.

Chapter 8- The application of biotechnology for industrial purposes is known as industrial biotechnology. It makes use of microorganisms and enzymes for the production of goods such as plastics, food and chemicals. The topics elaborated in this chapter will help in gaining a better perspective about the branches and applications of industrial biotechnology.

Finally, I would like to thank the entire team involved in the inception of this book for their valuable time and contribution. This book would not have been possible without their efforts. I would also like to thank my friends and family for their constant support.

Jordan Goldberg

Biotechnology: An Introduction

The field of biology that makes use of living systems and organisms for developing and making products is known as biotechnology. The branches of biotechnology include medical, animal, agricultural, environmental and industrial biotechnology. This is an introductory chapter which will introduce briefly all the significant aspects of these branches of biotechnology.

Biological Engineering

Biological Engineering is an interdisciplinary area focusing on the application of engineering principles to analyze biological systems and to solve problems in the interfacing of such systems - plant, animal or microbial with human-designed machines, structures, processes and instrumentation. The biological revolution continues to mature and impact all of us. Human-based gene manipulation affects nearly all North American food supplies. Plants and animals are already being defined on a molecular basis. Living organisms can now be analyzed, measured and "engineered" as never before. Designer "bugs" are being produced to enhance biological processes. These changes continue to redefine our research and graduate programs that continue to emphasize biological, environmental and food and fiber engineering. Our connections to agriculture and food systems remain, but modern agriculture is greatly influenced by biotechnology, and our connections to agriculture reflect this fact. A basic goal is to design technology that operates in harmony with the biology of living systems.

Biotechnology

Contrary to its name, biotechnology is not a single technology. Rather it is a group of technologies that share two (common) characteristics - working with living cells and their molecules and having a wide range of practice uses that can improve our lives.

Biotechnology can be broadly defined as "using organisms or their products for commercial purposes." As such, (traditional) biotechnology has been practices since he beginning of records history. (It has been used to:) bake bread, brew alcoholic beverages, and breed food crops or domestic animals. But recent developments in molecular biology have given biotechnology new meaning, new prominence, and new potential. It is (modern) biotechnology that has captured the attention of the public. Modern biotechnology can have a dramatic effect on the world economy and society.

One example of modern biotechnology is genetic engineering. Genetic engineering is the process of transferring individual genes between organisms or modifying the genes in an organism to remove or add a desired trait or characteristic.

Through genetic engineering, genetically modified crops or organisms are formed. These GM cr-ops or GMOs are used to produce biotech-derived foods. It is this specific type of modern biotech-nology, genetic engineering, that seems to generate the most attention and concern by consumers and consumer groups. What is interesting is that modern biotechnology is far more precise than traditional forms of biotechnology and so is viewed by some as being far safer.

Working of Modern Biotechnology

All organisms are made up of cells that are programmed by the same basic genetic material, called DNA (deoxyribonucleic acid). Each unit of DNA is made up of a combination of the following nucleotides - adenine (A), guanine (G), thymine (T), and cytosine (D) - as well as a sugar and a phosphate. These nucleotides pair up into strands that twist together into a spiral structure call a "double helix." This double helix is DNA. Segments of the DNA tell individual cells how to produce specific proteins. These segments are genes. It is the presence or absence of the specific protein that gives an organism a trait or characteristic. More than 10,000 different genes are found in most plant and animal species. This total set of genes for an organism is organized into chromosomes within the cell nucleus. The process by which a multicellular organism develops from a single cell through an embryo stage into an adult is ultimately controlled by the genetic information of the cell, as well as interaction of genes and gene products with environmental factors.

When cells reproduce, the DNA strands of the double helix separate. Because nucleotide A al-ways pairs with T and G always pairs with C, each DNA strand serves as a precise blueprint for a specific protein. Except for mutations or mistakes in the replication process, a single cell is equipped with the information to replicate into millions of identical cells. Because all organisms are made up of the same type of genetic material (nucleotides A, T, G, and C), biotechnologists use enzymes to cut and remove DNA segments from one organism and recombine it with DNA in another organism. This is called recombinant DNA (rDNA) technology, and it is one of the basic tools of modern biotechnology. rDNA technology is the laboratory manipulation of DNA in which DNA, or fragments of DNA from different sources, are cut and recombined using enzymes. This recombinant DNA is then inserted into a living organism. rDNA technology is usually used synonymously with genetic engineering. rDNA technology allows researchers to move genetic information between unrelated organisms to produce desired products or characteristics or to eliminate undesirable characteristics.

Genetic engineering is the technique of removing, modifying or adding genes to a DNA molecule in order to change the information it contains. By changing this information, genetic engineering changes the type or amount of proteins an organism is capable of producing. Genetic engineering is used in the production of drugs, human gene therapy, and the development of improved plants. For example, an "insect protection" gene (Bt) has been inserted into several crops - corn, cotton, and potatoes - to give farmers new tools for integrated pest management. Bt corn is resistant to European corn borer. This inherent resistance thus reduces a farmers pesticide use for controlling European corn borer, and in turn requires less chemicals and potentially provides higher yielding Agricultural Biotechnology.

Although major genetic improvements have been made in crops, progress in conventional breed-ing programs has been slow. In fact, most crops grown in the US produce less than their full ge-netic potential. These shortfalls in yield are due to the inability of crops to tolerate or adapt to from the child's DNA fingerprint; what remains comes from the biological father. These segments are then compared for a match with the DNA fingerprint of the alleged father.

DNA testing is also used on human fossils to determine how closely related fossil samples are from different geographic locations and geologic areas. The results shed light on the history of human evolution and the manner in which human ancestors settled different parts of the world.

Biotechnology for the 21st Century

Experts in United States anticipate the world's population in 2050 to be approximately 8.7 billion persons. The world's population is growing, but its surface area is not. Compounding the effects of population growth is the fact that most of the earth's ideal farming land is already being utilized. To avoid damaging environmentally sensitive areas, such as rain forests, we need to increase crop yields for land currently in use. By increasing crop yields, through the use of biotechnology the constant need to clear more land for growing food is reduced.

Countries in Asia, Africa, and elsewhere are grappling with how to continue feeding a growing population. They are also trying to benefit more from their existing resources. Biotechnology holds the key to increasing the yield of staple crops by allowing farmers to reap bigger harvests from currently cultivated land, while preserving the land's ability to support continued farming.

Malnutrition in underdeveloped countries is also being combated with biotechnology. The Rockefeller Foundation is sponsoring research on "golden rice", a crop designed to improve nutrition in the developing world. Rice breeders are using biotechnology to build Vitamin A into the rice. Vitamin A deficiency is a common problem in poor countries. A second phase of the project will increase the iron content in rice to combat anemia, which is widespread problem among women and children in underdeveloped countries. Golden rice, expected to be for sale in Asia in less than five years, will offer dramatic improvements in nutrition and health for millions of people, with little additional costs to consumers.

Similar initiatives using genetic manipulation are aimed at making crops more productive by reducing their dependence on pesticides, fertilizers and irrigation, or by increasing their resistance to plant diseases. Increased crop yield, greater flexibility in growing environments, less use of chemical pesticides and improved nutritional content make agricultural biotechnology, quite literally, the future of the world's food supply.

Concerns about Biotechnology

As biotechnology has become widely used, questions and concerns have also been raised. The most vocal opposition has come from European countries. One of the main areas of concern is the safety of genetically engineered food.

In assessing the benefits and risks involved in the use of modern biotechnology, there are a series of issues to be addressed so that informed decisions can be made. In making value judgments about risks and benefits in the use of biotechnology, it is important to distinguish between technology-inherent risks and technology-transcending risks. The former includes assessing any risks associated with food safety and the behavior of a biotechnology-based product in the environment. The latter involve the political and social context in which the technology is used, including how these uses may benefit or harm the interests of different groups in society. The health effects of foods grown from genetically engineered crop depend on the composition of the food itself. Any new product may have either beneficial or occasional harmful effects on human health. For example, a biotech-derived food with a higher content of digestible iron is likely to have a positive effect if consumed by iron-deficient individuals. Alternatively, the transfer of genes from one species to another may also transfer the risk for exposure to allergens. These risks are systematically evaluated by FDA and identified prior to commercialization. Individuals allergic to certain nuts, for example, need to know if genes conveying this trait are transferred to other foods such as soybeans. Labeling would be required if such crops were available to consumers.

Among the potential ecological risks identified are increased weediness, due to cross- pollination from genetically modified crops spreads to other plants in nearby fields. This may allow the spread of traits such as herbicide-resistance to non-target plants that could potentially develop into weeds. This ecological risk is assessed when deciding if a plant with a given trait should be released into a particular environment, and if so, under what conditions.

Other potential ecological risks stem from the use of genetically modified corn and cotton with insecticidal genes from Bacillus thuringiensis (Bt genes). This may lead to the development of resistance to Bt in insect populations exposed to the biotech-derived crop. There also may be risks to non-target species, such as birds and butterflies, from the plants with Bt genes. It is Important to keep all risks in perspective by comparing the products of biotechnology and conventional agriculture.

The reduction of biodiversity would represent a technology-transcending risk. Reduced biological diversity due to destruction of tropical forests, conversion of land to agriculture, overfishing, and the other practices to feed a growing world population is a significant loss far more than any potential loss of biodiversity due to biotech-derived crop varieties. Improved governance and international support are necessary to limit loss of biodiversity. What we know from our understanding of science and more than a decade of experience with biotech-derived plants is the following: There is no evidence that genetic transfers between unrelated organisms pose human health concerns that are different from those encountered with any new plant or animal variety. The risks associated with biotechnology are the same as those associated with plants and microbes developed by conventional methods.

Branches of Biotechnology

Based on applications, there are five main branches of biotechnology. They are:

- Animal Biotechnology: It deals with the development of transgenic animals for increased milk or meat production with resistance to various diseases. It also deals with in vitro fertilization of egg and transfer of embryo to the womb of female animal for further development.

- Medical Biotechnology: It deals with diagnosis of various diseases; large scale production of various drugs and hormones such as human insulin and interferon; vaccines for chicken pox, rabies, polio etc., and growth hormones, such as bovine. In the field of medical science, genetic engineering has helped in the large scale production of hormones, blood serum proteins; in the development of antibiotics, and other medically useful products.

- Industrial Biotechnology: It deals with commercial production of various useful organic substances, such as acetic acid, citric acid, acetone, glycerine, etc., and antibiotics like penicillin, streptomycin, mitomycin, etc., through the use of microorganisms especially fungi and bacteria.

- Environmental Biotechnology: It deals with detoxification of waste and industrial effluents, treatment of sewage water, and control of plant diseases and insects through the use of biological agents, such as viruses, bacteria, fungi etc.

- Plant Biotechnology: Plant Biotechnology is a combination of tissue culture and genetic engineering. It deals with development of transgenic plants with resistance to biotic and abiotic stress; development of haploids, embryo rescue, clonal multiplication, cryopreservation etc.

Tools of Biotechnology

The following are some important tools of biotechnology.

Restriction Enzymes

Restriction enzymes, also known as 'restriction endonucleases' are molecular scissors that can cut DNA at specific locations. These are part of a larger class of enzymes called 'Nucleases'. Nucleases are of two kinds:

- Exonucleases – Enzymes that remove nucleotides from the ends of DNA.

- Endonucleases – Enzymes that make cuts at specific positions in the DNA.

The first restriction endonuclease to be isolated, identifies a specific six base pair sequence and always cuts the DNA at a particular point. This specific sequence is the 'Recognition sequence'. Today, there are more than 900 restriction enzymes, isolated from about 230 bacterial strains. Each of these enzymes recognizes different recognition sequences. There is a naming convention to name restriction enzymes. Let's learn this using EcoRI as an example:

- Eco – indicates that this enzyme was isolated from Escherichia coli.

- R – refers to the name of the strain.

- I – is the Roman numeral that indicates the order of isolation of this restriction enzyme from the strain of bacteria.

Mode of Action of Restriction Enzymes

Endonucleases perform their function in the following manner:

- Inspect the length of DNA for the specific recognition sequence.

- Bind the DNA at the recognition sequence.

- Cut the two strands of DNA at specific points in their sugar-phosphate backbone.

Restriction enzymes recognize specific palindromic nucleotide sequences in the DNA. What is a palindrome? In terms of words, a palindrome is a group of letters that produce the same word when reading forward or backwards. An example is 'RADAR'. In terms of DNA, a palindrome is a sequence of base pairs that reads the same on both DNA strands when reading in the same orientation. For example:

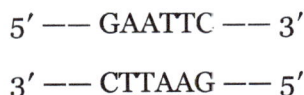

5' —— GAATTC —— 3'

3' —— CTTAAG —— 5'

On reading both the strands shown above in the 5' to 3' direction, they give the same sequence. This is true even when they are both read in the 3' to 5' direction.

Restriction enzymes cut the DNA strand a little away from the center of the palindromic site, but between the same two bases on both strands. This gives rise to single-stranded, overhanging stretches on each strand called 'sticky ends'. They are called 'sticky' because they can bind to their complementary cut counterparts.

Using restriction enzymes we can generate recombinant DNA that contains DNA from different sources. Cutting these DNA sources with the same restriction enzyme gives fragments with the same kind of 'sticky ends', that can then be joined using DNA ligases.

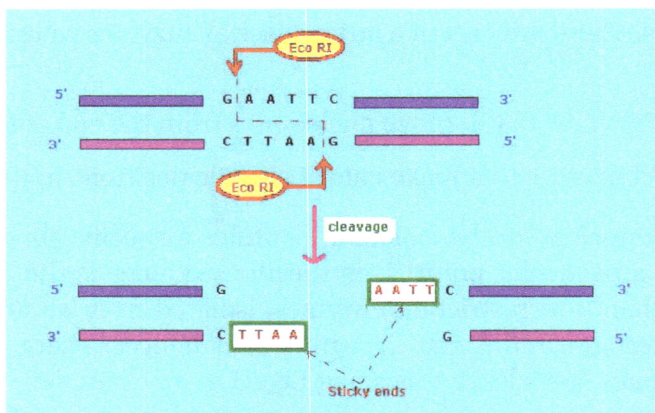

Mode of action of EcoRI.

Cloning Vectors

Just the way a mosquito acts as a 'vector' to transfer the malarial parasite into the human body, we need vectors to transfer the cut DNA into a host organism. Vectors are one of the important tools of biotechnology.

Plasmids make good vectors because they can replicate in bacterial cells, independent of the control of the chromosomal DNA. The vectors in use currently are engineered such that they help in easy linking of foreign DNA and allow selection of recombinants over non-recombinants. A vector needs the following features to enable cloning:

(i) Origin of Replication (ori): This is the sequence from where replication begins. Linking a piece of DNA to this sequence causes it to replicate in the host cell. This sequence also controls the copy number of the linked DNA. Therefore, the target DNA needs to be cloned into a vector whose 'ori' supports high copy number, in order to recover large amounts of the DNA.

(ii) Selectable Marker: The vector also needs to have a selectable marker which allows the selection of recombinants over non-recombinants. In terms of E. coli, some useful selectable markers are genes that provide resistance to antibiotics like ampicillin, kanamycin, chloramphenicol etc. Since the normal E. coli cells do not carry these resistance genes, it becomes easy to select the recombinants.

(iii) Cloning Sites: In order to attach the foreign DNA to a vector, the vector should have a recognition site for a specific restriction enzyme. Multiple recognition sites will result in multiple DNA fragments, complicating the process of cloning. A vector has more than one antibiotic resistance gene. The foreign DNA is ligated into a restriction site in one of the antibiotic resistance genes.

For example, let's say an E. coli cloning vector has genes for ampicillin and tetracycline resistance. On ligating the foreign DNA into a recognition site within the tetracycline resistance gene, the plasmid loses its tetracycline resistance. But, it can still be selected for non-recombinants by plating on ampicillin-containing medium.

Now, by transferring the ones that grow in the ampicillin medium to a medium with tetracycline, we can dissect out recombinants from non-recombinants. The recombinants will grow in ampicillin but not in tetracycline medium; while non-recombinants will grow in both mediums.

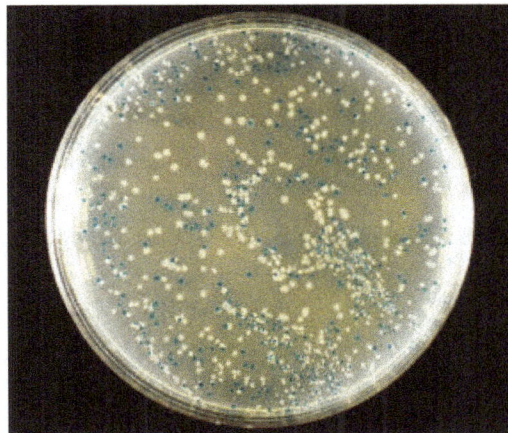

Blue-white screening using B-galactosidase.

Currently, alternative markers are available that can differentiate recombinants from non-recombinants based on their ability to produce color. This involves insertion of the foreign DNA in the DNA sequence of an enzyme like β-galactosidase, which inactivates the enzyme. This is 'insertional inactivation'. On reaction with a substrate, the recombinants do not produce color whereas non-recombinants produce color.

(iv) Vectors to Clone Genes in Plants and Animals: Long before us, bacteria and viruses knew how to transfer genes into plants and animals. For example, Agrobacterium tumifaciens, a pathogen on dicot plants, transfers 'T-DNA' that transforms normal plants cells into tumours. These tumours then produce chemicals that the pathogen requires. With better understanding, we have now converted these pathogens into useful vectors to deliver genes of interest to plants or animals.

Now, the tumour-inducing (Ti) plasmid of Agrobacterium tumifaciens has been modified into a cloning vector. This vector is no longer harmful but is useful in delivering genes of interest to plants. Retroviruses transform normal cells into cancerous cells in animals. These have also been modified so that they are no longer harmful and can deliver genes to animals.

Competent Host

Host cells are bacterial cells which take up the recombinant DNA. Since DNA is hydrophilic, it cannot pass through the cell membrane of bacteria easily. Therefore, the bacterial cells have to be made 'competent' to take up the DNA.

Some procedures that make the cells competent are treatments with a specific concentration of divalent cation like calcium. This makes it easy for the DNA to enter the cell wall through pores. Incubation of cells with the recombinant DNA on the ice, followed by heat shock at 42 °C and another incubation on ice, enables the cells to take up the DNA.

There are several other methods to introduce foreign DNA into host cells. The 'microinjection' method involves injecting the recombinant DNA directly into the nucleus of an animal cell. The 'bolistics' or 'gene gun' method bombards plant cells with high-velocity microparticles of gold or tungsten coated with DNA. The last method uses 'disarmed pathogen vectors' to transfer the recombinant DNA into the infected host cells.

References

- Biological-engineering, research: cornell.edu, Retrieved 15 June, 2019

- Biotech-applications, documents, extension-program: ncsu.edu, Retrieved 21 April, 2019

- Biotechnology-scope-and-branches-of-biotechnology, branches-biotechnology, biotechnology: biologydiscussion.com, Retrieved 15 January, 2019

- Tools-of-biotechnology, biotechnology-principles-and-process, biology, guides: toppr.com, Retrieved 21 July, 2019

Recombinant DNA Technology: Processes and Applications

Any piece of DNA which has been created by combining at least two strands is known as recombinant DNA. They are formed using the methods of genetic recombination for the purpose of combining genetic material from different sources to create sequences which are not found in the genome. This chapter discusses in detail the processes and applications related to recombinant DNA technology such as DNA cloning and DNA replication.

Recombinant DNA

Recombinant DNA refers to the molecules of DNA from two different species that are inserted into a host organism to produce new genetic combinations that are of value to science, medicine, agriculture, and industry. Since the focus of all genetics is the gene, the fundamental goal of laboratory geneticists is to isolate, characterize, and manipulate genes. Although it is relatively easy to isolate a sample of DNA from a collection of cells, finding a specific gene within this DNA sample can be compared to finding a needle in a haystack. Consider the fact that each human cell contains approximately 2 metres (6 feet) of DNA. Therefore, a small tissue sample will contain many kilometres of DNA. However, recombinant DNA technology has made it possible to isolate one gene or any other segment of DNA, enabling researchers to determine its nucleotide sequence, study its transcripts, mutate it in highly specific ways, and reinsert the modified sequence into a living organism.

In Vitro Mutagenesis

Another use of cloned DNA is in vitro mutagenesis in which a mutation is produced in a segment of cloned DNA. The DNA is then inserted into a cell or organism, and the effects of the mutation are studied. Mutations are useful to geneticists in enabling them to investigate the components of any biological process. However, traditional mutational analysis relied on the occurrence of random spontaneous mutations—a hit-or-miss method in which it was impossible to predict the precise type or position of the mutations obtained. In vitro mutagenesis, however, allows specific mutations to be tailored for type and for position within the gene. A cloned gene is treated in the test tube (in vitro) to obtain the specific mutation desired, and then this fragment is reintroduced into the living cell, where it replaces the resident gene.

One method of in vitro mutagenesis is oligonucleotide-directed mutagenesis. A specific point in a sequenced gene is pinpointed for mutation. An oligonucleotide, a short stretch of synthetic DNA of the desired sequence, is made chemically. For example, the oligonucleotide might have adenine in one specific location instead of guanine. This oligonucleotide is hybridized to the complementary strand of the cloned gene; it will hybridize despite the one base pair mismatch. Various enzymes are added to allow the oligonucleotide to prime the synthesis of a complete strand within the vector. When the vector is introduced into a bacterial cell and replicates, the mutated strand will act as a template for a complementary strand that will also be mutant, and thus a fully mutant molecule is obtained. This fully mutant cloned molecule is then reintroduced into the donor organism, and the mutant DNA replaces the resident gene.

Another version of in vitro mutagenesis is gene disruption, or gene knockout. Here, the resident functional gene is replaced by a completely nonfunctional copy. The advantage of this technique over random mutagenesis is that specific genes can be knocked out at will, leaving all other genes untouched by the mutagenic procedure.

In gene knockout a functional gene is replaced by an inactivated gene that is created using recombinant DNA technology. When a gene is "knocked out," the resulting mutant phenotype (observable characteristics) often reveals the gene's biological function.

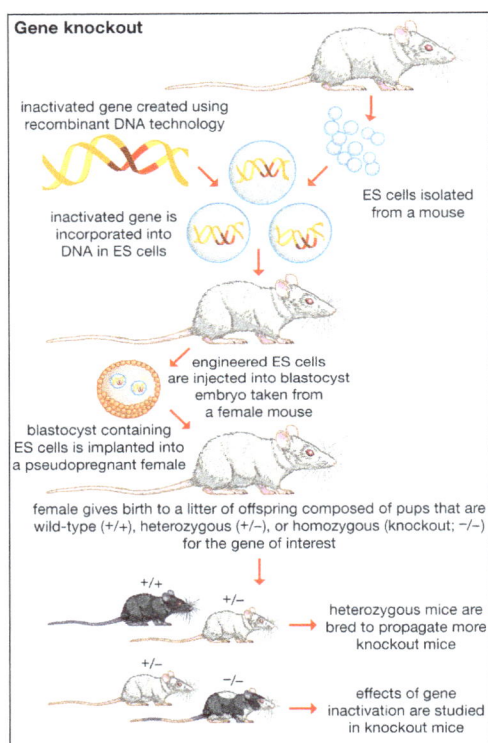

In gene knockout a functional gene is replaced by an inactivated gene that is created using recombinant DNA technology. When a gene is "knocked out," the resulting mutant phenotype (observable characteristics) often reveals the gene's biological function.

Genetically Modified Organisms

The ability to obtain specific DNA clones using recombinant DNA technology has made it possible to add the DNA of one organism to the genome of another. The added gene is called a transgene. The transgene inserts itself into a chromosome and is passed to the progeny as a new component of the genome. The resulting organism carrying the transgene is called a transgenic organism or a genetically modified organism (GMO). In this way, a "designer organism" is made that contains some specific change required for an experiment in basic genetics or for improvement of some commercial strain. Several transgenic plants have been produced. Genes for toxins that kill insects have been introduced in several species, including corn and cotton. Bacterial genes that confer resistance to herbicides also have been introduced into crop plants. Other plant transgenes aim at improving the nutritional value of the plant.

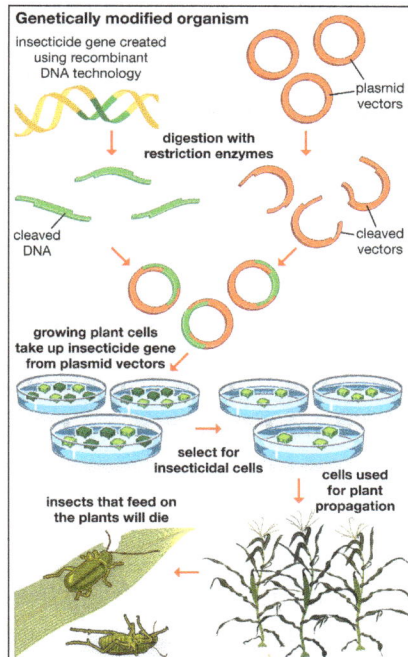

Genetically modified organisms are produced using scientific methods that include recombinant DNA technology.

Genetically modified corn (maize).

Gene Therapy

Gene therapy is the introduction of a normal gene into an individual's genome in order to repair a mutation that causes a genetic disease. When a normal gene is inserted into a mutant nucleus, it most likely will integrate into a chromosomal site different from the defective allele; although this may repair the mutation, a new mutation may result if the normal gene integrates into another functional gene. If the normal gene replaces the mutant allele, there is a chance that the transformed cells will proliferate and produce enough normal gene product for the entire body to be restored to the undiseased phenotype. So far, human gene therapy has been attempted only on somatic (body) cells for diseases such as cancer and severe combined immunodeficiency syndrome (SCIDS). Somatic cells cured by gene therapy may reverse the symptoms of disease in the treated individual, but the modification is not passed on to the next generation. Germinal gene therapy aims to place corrected cells inside the germ line (e.g., cells of the ovary or testis). If this is achieved, these cells will undergo meiosis and provide a normal gametic contribution to the next generation. Germinal gene therapy has been achieved experimentally in animals but not in humans.

Reverse Genetics

Recombinant DNA technology has made possible a type of genetics called reverse genetics. Traditionally, genetic research starts with a mutant phenotype, and, by Mendelian crossing analysis, a researcher is able to attribute the phenotype to a specific gene. Reverse genetics travels in precisely the opposite direction. Researchers begin with a gene of unknown function and use molecular analysis to determine its phenotype. One important tool in reverse genetics is gene knockout. By mutating the cloned gene of unknown function and using it to replace the resident copy or copies, the resultant mutant phenotype will show which biological function this gene normally controls.

Diagnostics

Recombinant DNA technology has led to powerful diagnostic procedures useful in both medicine and forensics. In medicine these diagnostic procedures are used in counseling prospective parents as to the likelihood of having a child with a particular disease, and they are also used in the prenatal prediction of genetic disease in the fetus. Researchers look for specific DNA fragments that are located in close proximity to the gene that causes the disease of concern. These fragments, called restriction fragment length polymorphisms (RFLPs), often serve as effective "genetic markers." In forensics, DNA fragments called variable number tandem repeats (VN-TRs), which are highly variable between individuals, are employed to produce what is called a "DNA fingerprint." A DNA fingerprint can be used to determine if blood or other body fluids left at the scene of a crime belongs to a suspect.

Genomics

The genetic analysis of entire genomes is called genomics. Such a broadscale analysis has been made possible by the development of recombinant DNA technology. In humans, knowledge of the entire genome sequence has facilitated searching for genes that produce hereditary diseases. It is also capable of revealing a set of proteins—produced at specific times, in specific tissues, or in specific diseases—that might be targets for therapeutic drugs. Genomics also allows the comparison of one genome with another, leading to insights into possible evolutionary relationships between organisms.

Genomics has two subdivisions: structural genomics and functional genomics. Structural genomics is based on the complete nucleotide sequence of a genome. Each member of a library of clones is physically manipulated by robots and sequenced by automatic sequencing machines, enabling a very high throughput of DNA. The resulting sequences are then assembled by a computer into a complete sequence for every chromosome. The complete DNA sequence is scanned by computer to find the positions of open reading frames (ORFs), or prospective genes. The sequences are then compared to the sequences of known genes from other organisms, and possible functions are assigned. Some ORFs remain unassigned, awaiting further research.

Functional genomics attempts to understand function at the broadest level (the genomic level). In one approach, gene functions of as many ORFs as possible are assigned as above in an attempt to obtain a full set of proteins encoded by the genome (called a proteome). The proteome broadly defines all the cellular functions used by the organism. Function in relation to specific developmental stages also is assessed by trying to identify the "transcriptome," the set of mRNA transcripts made at specific developmental stages. The practical approach utilizes microarrays—glass plates the size of a microscope slide imprinted with tens of thousands of ordered DNA samples, each representing one gene (either a clone or a synthesized segment). The mRNA preparation under test is labeled with a fluorescent dye, and the microarray is bathed in this mRNA. Fluorescent spots appear on the array indicating which mRNAs were present, thus defining the transcriptome.

Protein Manufacture

Recombinant DNA procedures have been used to convert bacteria into "factories" for the synthesis of foreign proteins. This technique is useful not only for preparing large amounts of protein for basic research but also for producing valuable proteins for medical use. For example, the genes for human proteins such as growth hormone, insulin, and blood-clotting factor can be commercially manufactured. Another approach to producing proteins via recombinant DNA technology is to introduce the desired gene into the genome of an animal, engineered in such a way that the protein is secreted in the animal's milk, facilitating harvesting.

Steps in Recombinant DNA Technology

Basic steps involved in Recombinant DNA technology are:

1. Selection and isolation of DNA insert:

    First step in rec DNA technology is the selection of a DNA segment of interest which is to be cloned. This desired DNA segment is then isolated enzymatically. This DNA segment of interest is termed as DNA insert or foreign DNA or target DNA or cloned DNA.

2. Selection of suitable cloning vector:

    A cloning vector is a self-replicating DNA molecule, into which the DNA insert is to be integrated. A suitable cloning vector is selected in the next step of rec DNA technology. Most commonly used vectors are plasmids and bacteriophages.

3. Introduction of DNA-insert into vector to form recDNA molecule:

    The target DNA or the DNA insert which has been extracted and cleaved enzymatically by the selective restriction endonuclease enzymes are now ligated (joined) by the enzyme ligase to vector DNA to form a rec DNA molecule which is often called as cloning-vectoring-vector-insert DNA construct.

4. recDNA molecule is introduced into a suitable host:

    Suitable host cells are selected and the rec DNA molecule so formed is introduced into these host cells. This process of entry of rec DNA into the host cell is called transformation. Usually selected hosts are bacterial cells like E. coli, however yeast, fungi may also be utilized.

5. Selection of transformed host cells:

    Transformed cells (or recombinant cells) are those host cells which have taken up the recD-NA molecule. In this step the transformed cells are separated from the non-transformed cells by using various methods making use of marker genes.

6. Expression and Multiplication of DNA insert in the host:

Finally, it is to be ensured that the foreign DNA inserted into the vector DNA is expressing the desired character in the host cells. Also, the transformed host cells are multiplied to obtain sufficient number of copies. If needed, such genes may also be transferred and expressed into another organism.

Tools for Recombinant DNA Technology

WDNA technology utilizes a number of biological tools to achieve its objectives, most important of them being the enzymes.

Important biological tools for rec DNA technology are:

1. Enzymes:
    a. Restriction Endonucleases
    b. Exonucleases
    c. DNA ligases
    d. DNA polymerase
2. Cloning Vector,
3. Host organism,
4. DNA insert or foreign DNA,
5. Linker and adaptor sequences.

An account of all these biological tools of genetic engineering is given below.

Enzymes

A number of specific enzymes are utilized to achieve the objectives of rec DNA technology. The enzymology of genetic engineering includes the following types of enzymes.

Restriction Endonuclease

These enzymes serve as important tools to cut DNA molecules at specific sites, which is the basic need for rec DNA technology.

These are the enzymes that produce internal cuts (cleavage) in the strands of DNA, only within or near some specific sites called recognition sites/recognition sequences/ restriction sites 01 target sites. Such recognition sequences are specific for each restriction enzyme. Restriction endonuclease enzymes are the first necessity for rec DNA technology.

The presence of restriction enzymes was first of all reported by W. Arber in the year 1962. He found that when the DNA of a phage was introduced into a host bacterium, it was fragmented into small pieces. This led him to postulate the presence of restriction enzymes. The first true restriction endonuclease was isolated in 1970s from the bacterium E. coli by Meselson and Yuan.

Another important breakthrough was the discovery of restriction enzyme. They isolated it from -the bacterium Haemophilus influenza. In the year 1978, the Nobel Prize for Physiology and Medicine was given to Smith, Arber and Nathans for the discovery of endonucleases.

Types of Restriction Endonucleases

There are 3 main categories of restriction endonuclease enzymes:

- Type-I Restriction Endonucleases.

- Type-II Restriction Endonucleases.

- Type-III Restriction Endonucleases.

Type-I Restriction Endonucleases

These are the complex type of endonucleases which cleave only one strand of DNA. These enzymes have the recognition sequences of about 15 bp length.

They require Mg^{++} ions and ATP for their functioning. Such types of restriction endonucleases cleave the DNA about 1000 bp away from the 5′ end of the sequence 'TCA' located within the recognition site. Important examples of Type-I restriction endonuclease enzyme are EcoK, EcoB, etc.

Table: Recognition sequences of several Restriction Endonucleases.

Enzyme	Source organism	Recognition Sequence 5′....3′	Blunt or sticky ends.
EcoR l	Escherichia coli	GAATCC	Sticky
Bgil ll	Bacillus globigii	AGATCT	Sticky
Hind ll	Haemophilus influenzae	GTPyPuAC	Blunt
Hind lll	Haemophilus influenzae	AAGCTT	Sticky
Hinfl	Haemophilus influenzae	GANTC	Sticky
Hpa l	H.parainfluenzae	GTTAAC	Blunt
Hae lll	H.aegyptius	GGCC	Blunt
Sau 3A	Staphylococcus aureus	GATC	Sticky
Pvu l	Proteus vulgaris	CGATCG	Sticky
BamH l	Bacillus amyloliquegaciens	GGATCC	Sticky
Taq l	Thermus aquaticus	TCGA	Sticky
Smal	Serratia marcescens	CCCGGG	Blunt
Sfi l	Streptomyces fimbriatus	GGCCNNNNNGGCC	Sticky
Sal l	Streptomyces albus	GTCGAC	Sticky

Type-II Restriction Endonucleases

These are most important endonucleases for gene cloning and hence for rec DNA technology. These enzymes are most stable. They show cleavage only at specific sites and therefore they produce the DNA fragments of a defined length. These enzymes show cleavage in both the strands of DNA, immediately outs.de then- recognition sequences. They require Mg^{++} ions for their functioning.

Such enzymes are advantageous because they don't require ATP for cleavage and they cause cleavage in both strands of DNA. Only Type II Restriction Endonucleases are used tor gene cloning due to their suitability.

The recognition sequences for Type-II Restriction Endonuclease enzymes are in the form of palindromic sequences with rotational symmetry, i.e., the base sequence .n the first half of one strand of DNA is the mirror image of the second half of other strand of that DNA double helix. Important examples of Type-II Restriction endonucleases include Hinfl, EcoRI, PvuII, Alul, Haelll etc.

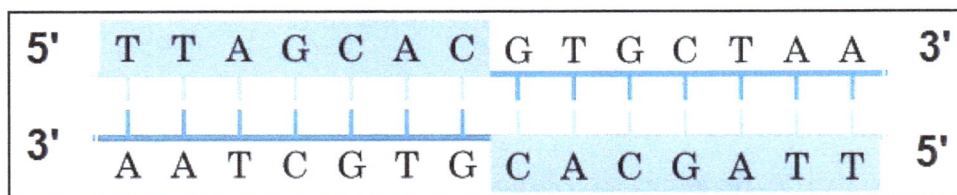

A palindrome with rotational symmetry.

Type-III Restriction Endonucleases

These are not used for gene cloning. They are the intermediate enzymes between Type-I and Type-II restriction endonuclease. They require Mg^{++} ions and ATP for cleavage and they cleave the DNA at well-defined sites in the immediate vicinity of recognition sequences, e.g. Hinf III, etc.

Nature of Cleavage by Restriction Endonucleases

The nature of cleavage produced by a restriction endonuclease is of considerable importance.

They cut the DNA molecule in two ways:

i.   Many restriction endonucleases cleave both strands of DNA simply at the same point within the recognition sequence. As a result of this type of cleavage, the DNA fragments with blunt ends are generated. PvuII, Haelll, Alul are the examples of restriction endonucleases producing blunt ends. Blunt ends may also be referred to as flush ends.

ii. In the other style of cleavage by the restriction endonucleases, the two strands of DNA are cut at two different points. Such cuts are termed as staggered cuts and this results into the generation of protruding ends i.e., one strand of the double helix extends a few bases beyond the other strand. Such ends are, called cohesive or sticky ends.

Such ends have the property to pair readily with each other when pairing conditions are provided. Another feature of the restriction endonucleases producing such sticky ends is that two or more of such enzymes with different recognition sequences may generate the same sticky ends.

Mode of Action Restriction Endonucleases.

Exonucleases

Exonuclease is an enzyme that removes nucleotides from the ends of a nucleic acid molecule. An exonuclease removes nucleotide from the 5′ or 3′ end of a DNA molecule. An exonuclease never produces internal cuts in DNA.

An Exonuclease activity: Nucleotides are removed from the end of DNA.

In rec DNA technology, various types of exonucleases are employed like Exonuclease Bal31, E. coli exonuclease III, Lambda exonuclease, etc.

Exonculease Bal31 are employed for making the DNA fragment with blunt ends shorter from both its ends.

E coli Exonuclease III is utilized for 3'end modifications because it has the capability to remove nucleotides from 3′-OH end of DNA.

Lambda exonuclease is used to modify 5′ ends of DNA as it removes the nucleotides from 5′ terminus of a linear DNA molecule.

DNA Ligase

The function of these enzymes is to join two fragments of DNA by synthesizing the phosphodiester bond. They function to repair the single stranded nicks in DNA double helix and in rec DNA technology they are employed for sealing the nicks between adjacent nucleotides. This enzyme is also termed as molecular glue.

DNA Polymerases

These are the enzymes which synthesize a new complementary DNA strand of an existing DNA or RNA template. A few important types of DNA polymerases are used routinely in genetic engineering. One such enzyme is DNA polymerase ! which , prepared from E coli. The Klenow fragment of DNA polymerase-I .s employed to make the protruding ends double-stranded by extension of the shorter strand.

Another type of DNA polymerase used in genetic engineering is Taq DNA polymerase which is used in PCR (Polymerase Chain Reaction).

Reverse transcriptase is also an important type of DNA polymerase enzyme for genetic engineering. It uses RNA as a template for synthesizing a new DNA strand called as cDNA a e complementary DNA). Its main use is in the formation of cDNA libraries. Apart from all these above mentioned enzymes, a few other enzymes also mark their importance in genetic engineering.

A brief description of these is given below:

(a) Terminal deoxynucleotidyl transferase enzyme: It adds single stranded sequences to 3′-terminus of the DNA molecule. One or more deoxynbonucleotides (dATP, dGTP, dl IP, dCTP) are added onto the 3′-end of the blunt-ended fragments.

(b) Alkaline Phosphatase Enzyme: It functions to remove the phosphate group from the 5′-end of a DNA molecule.

(c) Polynucleotide Kinase Enzyme: It has an effect reverse to that of Alkaline Phosphatase, i.e. it functions to add phosphate group to the 5′-terminus of a DNA molecule.

Cloning Vectors

It is another important natural tool which geneticists use in rec DNA technology. The cloning vector is the DNA molecule capable of replication in a host organism, into which the target DNA is introduced producing the rec DNA molecule.

A cloning vector may also be termed as a cloning vehicle or earner DNA or simply as a vector or a vehicle a great variety of cloning vectors are present for use with E. coli is the host organism.

However under certain circumstances it becomes desirable to use different host for cloning experiments. So, various cloning vectors have been developed based on other bacteria like Bacillus, Pseudomonas, Agrobacterium, etc. and on different eukaryotic organisms like yeast and other fungi.

The cloning vector which has only a single site for cutting by a particular restriction endormclease is Considered as a good cloning vector. Different types of DNA molecules may be used as cloning vehicles such as they may be plasmids, bacteriophages, cosmids, phasmids or artificial chromosomes.

Host Organism

A good host organism is an essential tool tor genetic engineering. Most widely used host for rec DNA technology is the bacterium E. coli. because cloning and isolation of DNA inserts is very easy in this host. A good host organism is the one winch easy to transform and in which the replication of rec DNA is easier. There should not be any interfering element against the replication of rec DNA in the host cells.

DNA Insert or Foreign DNA

The desired DNA segment which is to be cloned is called as DNA insert or foreign DNA or target DNA. The selection of a suitable target DNA is the very first step of rec DNA technology. The target DNA (gene) may be of viral, plant, animal or bacterial origin.

Following points must be kept in mind while selecting the foreign DNA:

1. CD It can be easily extracted from source.

2. It can be easily introduced into the vector.

3. The genes should be beneficial for commercial or research point of view.

A number of foreign genes are being cloned for benefit of human beings. Some of these DNA inserts are the genes responsible for the production of insulin, interferon's, lymphotoxins various growth factors, interleukins, etc.

Linker and Adaptor Sequences

Linkers and adaptors are the DNA molecules which help in the modifications of cut ends of DNA fragments. These can be joined to the cut ends and hence produce modifications as desired.

Both are short, chemically synthesized, double stranded DNA sequences. Linkers have (within them) one or more restriction endonuclease sites and adaptors have one or both sticky ends. Different types of linkers and adaptors are used for different purposes.

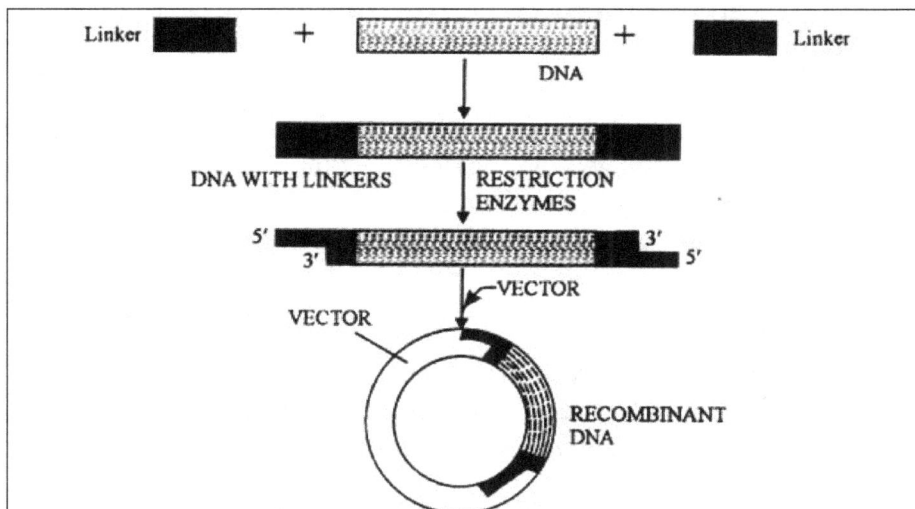

Modification of DNA ends by using linker during rec DNA technology.

Linkers contain target sites for the action of one or more restriction enzymes. They can be ligated to the blunt ends of foreign DNA or vector DNA. Then they undergo a treatment with a specific restriction endonuclease to produce cohesive ends of DNA fragments EcoRI-linker is a common example of frequently used linkers.

Adaptors are the chemically Synthesized molecules which have pre-formed cohesive ends. Adaptors are employed for end modification in cases where the recognition site for restriction endonuclease enzyme is present within the foreign DNA.

The foreign DNA is ligated with adaptor on both ends. This new molecule, so formed, is then phosphorylated at the 5'-terminii. Finally foreign DNA modified with adaptors is integrated into the vector DNA to form the recombinant DNA molecule.

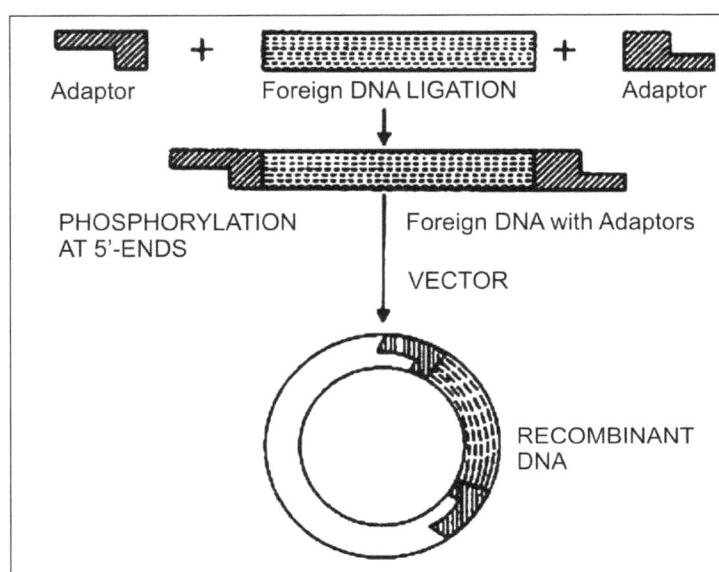

Construction of DNA Fragments with cohesive end using Adaptors.

Techniques used in Recombinant DNA Technology

A number of techniques are used for various purposes during different steps of rec DNA technology.

Such techniques serve for the fulfilment of different requirements or to obtain proper information for drawing an exact inference during genetic engineering. Some of these important techniques are gel electrophoresis, blotting techniques, dot-blot hybridization, DNA sequencing, artificial gene synthesis, polymerase chain reaction, colony hybridization, etc.

Gel Electrophoresis

It is the technique of separation of charged molecules (in aqueous phase) under the influence of an electrical field so that they move on the gel towards the electrode of opposite charge i.e., cations move towards the negative electrode and anions move towards the positive electrode.

The genomic DNA is extracted from the desired host and is then fragmented using restriction endonucleases.

For separation of these cut fragments and isolation of desired DNA fragment, the technique of gel electrophoresis is employed. Gel electrophoresis may be of horizontal or vertical type. Usually agarose gel is used for separation of large segments of DNA while the polyacrylamide gel is used for the separation of small DNA fragments which are only a few base pairs long.

Gel electrophoresis employs a buffer system, a medium which is a gel and a source of direct current. Samples having DNA fragments are applied on the gel and current is passed through the system for an appropriate time. Different DNA fragments move up to different distances on the gel depending on their charge to mass ratio.

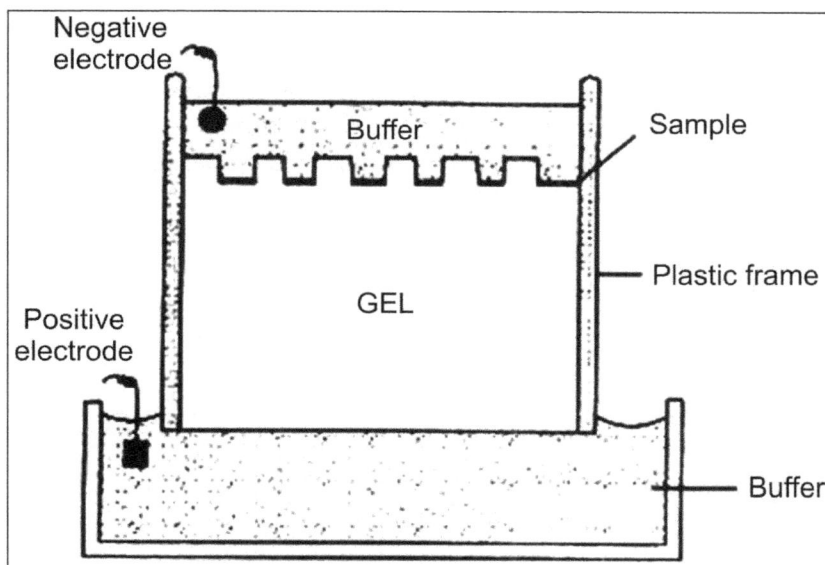

Vertical Gel Electrophoresis for Separation of DNA fragment.

The heavier fragments move a little, while the lighter DNA fragments move up to a larger distance. Following the migration of the molecules, the gel is treated with selective stains to show the location of separated molecules in the form of bands.

Very large DNA molecules or chromosomes cannot be separated even by Agarose Gel electrophoresis. For separation of such very large DNA molecules (sometimes representing whole chromosomes), a new technique is used which is known as Pulse Field Gel Electrophoresis (PFGE).

Blotting Techniques

Visualization of a specific DNA (or RNA or protein) fragment out of many molecules requires a technique called blot transfer. In this technique, the separated bands are transferred onto a nitrocellulose membrane from the gel.

Mainly there are three Types of Blot Transfer Procedures:

Southern Blotting, Northern Blotting and Western blotting.

Southern blotting is named after the person who devised this technique, viz. E.M. Southern. The other names began as laboratory jargon but they are now accepted terms.

Technically, blotting may be defined as the transfer of macromolecules from the gel onto the surface of an immobilizing membrane like nitrocellulose membrane. It is to note here that during such transfer, the relative positions of bands (of macromolecules) are same on the membrane as they occurred on the gel.

The membranes which may.be used in blotting are nitrocellulose membrane, nylon membrane, carboxymethyl membrane, diazobenzyl-oxymethyl (DBM) membranes, etc.

Southern blotting is used for the transfer of DNA from gel onto the membrane while Northern and Western blotting are used for the transfer of RNA and protein bands respectively. One other blotting technique is south-western blotting which examines the protein-DNA interactions.

In figure, a schematic representation of southern blotting technique is given in the. In this technique first of all, the sample DNA is digested with restriction enzymes to obtain fragments of different lengths. These differently sized DNA segments are then passed through Agarose Gel Electrophoresis for their separation based on their lengths.

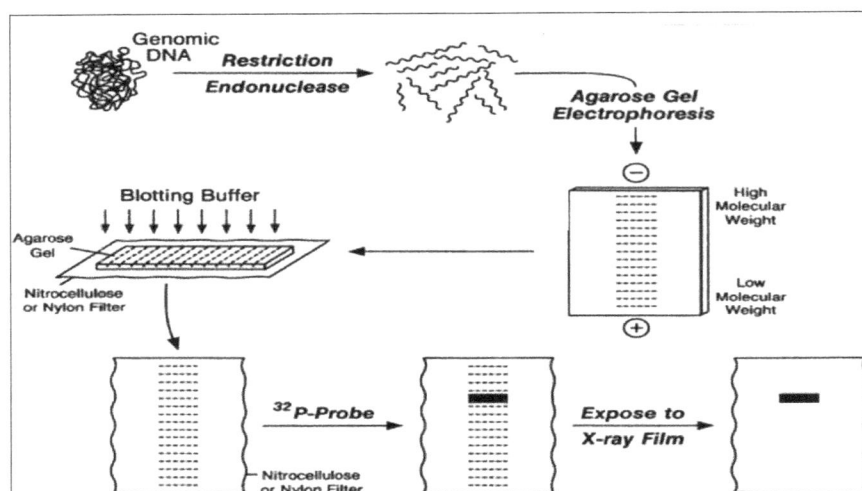

Method of Sourthen blotting for visualising DNA fragments.

The gel so obtained with different bands of DNA fragments is placed on top of buffer saturated filter papers which act as a filter paper wick. Above gel is put a nitrocellulose filter and over nitrocellulose filter are placed many dry filter paper sheets. With the movement of buffer towards the dry filter papers, the DNA bands are also moved upwards and hence they get bound to the nitrocellulose filter membrane.

Now, the nitrocellulose filter is removed and baked in vacuum. DNA fragments on the nitrocellulose filter are hybridized with single stranded radioactively labeled probes. Washing is done to remove unbound probes and finally the DNA bands with radioactivity are visualized by autoradiography.

In Northern Blotting, RNA molecules are blot transferred from the gel onto a chemically reactive paper. Western blotting is used for proteins and its working is based on the specificity of antibody-antigen reaction. In this technique the hybridization of bound proteins is done with radioactively labeled antibodies.

Dot Blot Hybridization

The procedure of this technique is almost the same as blotting, but the only difference is that the DNA fragments are not separated by electrophoresis, instead they are directly applied as a dot on the nitrocellulose membrane.

Then radioactively labeled DNA probes having the complementary base sequences to the DNA of interest are applied on this membrane to allow its hybridization. The position of this hybridization is then detected by autoradiography method.

Artificial Gene Synthesis

This technique may also be called as oligonucleotide synthesis. It is one of those techniques which have been adopted for the synthesis of desired gene or DNA fragment. Gene synthesis is now a routine laboratory procedure to be utilized in the rec DNA technology.

First success in the approach of artificial gene synthesis was achieved by Dr. Har Govind Khorana and his co-workers in 1970 when they synthesized the artificial gene for a t-RNA in vitro which had potential for functioning within a living cell.

Major Approaches Available for the Artificial Synthesis of Genes are as follows.

Enzymatic Synthesis of Gene

When details of base sequence of concerned gene are available, the polynucleotide of that same base sequence can be synthesized by enzymatic method. In this method the bacterial enzyme Polynucleotide phosphorylase is utilized. This method is easy to perform and does not require any template.

Chemical Synthesis of Gene

Once the base sequence of a gene is deducted, this gene can be synthesized by a purely chemical method as used by Khorana and his co-workers for the synthesis of gene for yeast alanyl t-RNA. This method utilizes different chemical reagents for various steps of the process.

There are mainly three distinct methods, which are phosphodiester phosphotriester, and phosphite-triester methods. These methods differ in their strategies for protecting the hydroxyl group of the phosphate residues.

If the detailed sequence of the concerned gene is unknown then the artificial gene is synthesized in the form of cDNA i.e. complementary DNA from the mRNA of that gene. In this method, the enzyme employed is RNA directed DNA polymerase.

Colony Hybridization Technique

This technique is used in genetic engineering for the identification of transformed bacterial cells (i.e. cells which contain foreign DNA). After transformation of cells with a specific DNA, it is likely that only some of those cells may have foreign DNA. For further procedure, firstly it is important to screen such cells which are having foreign DNA.

This screening is done by using the technique of colony hybridization in case of bacterial cells. A similar technique namely Plaque Hybridization is utilized for screening of transformed bacteriophages.

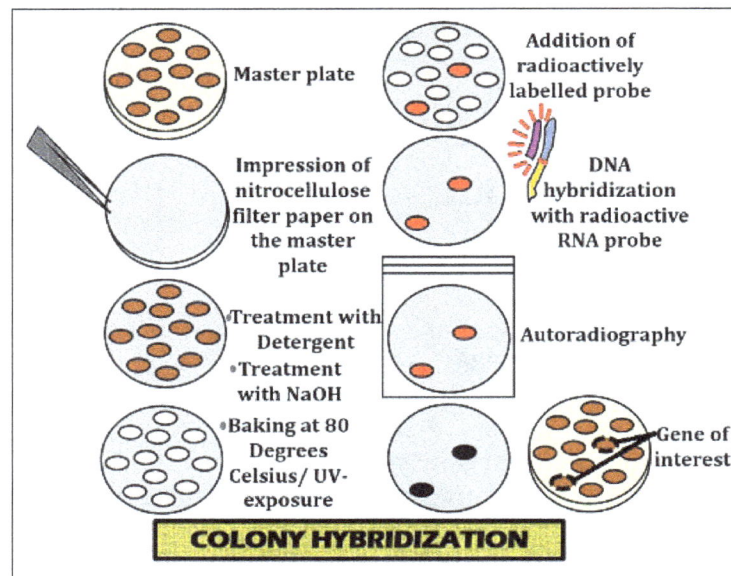

Steps in Colony Hybridization Technique.

Basic principle of this technique lies in the in-situ hybridization of transformed bacterial cells with a radioactive probe sequence. Due to the specificity of probe, it enables rapid identification of one colony (through radioactivity) even amongst many thousands of colonies.

The transformed bacterial cells are first of all plated on a suitable agar plate which is termed as the master plate. Colonies are grown in the master plate. These colonies on the master plate are replica-plated onto a nitrocellulose or nylon membrane by placing it gently over the master plate. This replica-plate carrying the colonies is removed and treated with alkaline reagent to lyse the bacteria.

DNA of those bacterial cells is denatured. Proteins on the membrane are digested. Finally the membrane is washed to remove all other molecules, leaving behind only the denatured DNA bound to it, in the form of DNA print of the colonies.

This DNA print is then hybridized with a radioactively labeled RNA/DNA probe. Membrane is washed to remove any unbound probe and then autoradiography is done to detect radioactivity. The positions of the DNA prints showing up in autoradiograph are then compared with the master plate to identify the transformed colony.

DNA Cloning

DNA Sequencing is the process of making multiple, identical copies of a particular piece of DNA. In a typical DNA cloning procedure, the gene or other DNA fragment of interest (perhaps a gene for a medically important human protein) is first inserted into a circular piece of DNA called a plasmid. The insertion is done using enzymes that "cut and paste" DNA, and it produces a molecule of recombinant DNA, or DNA assembled out of fragments from multiple sources.

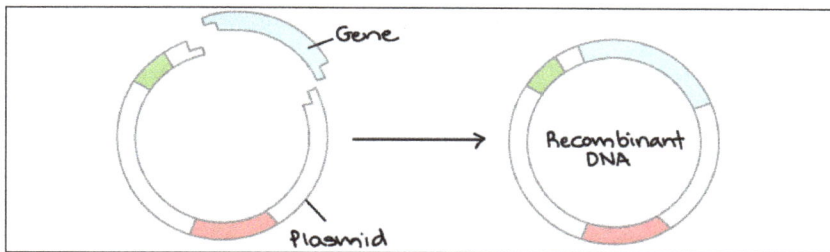

Next, the recombinant plasmid is introduced into bacteria. Bacteria carrying the plasmid are selected and grown up. As they reproduce, they replicate the plasmid and pass it on to their offspring, making copies of the DNA it contains.

What is the point of making many copies of a DNA sequence in a plasmid? In some cases, we need lots of DNA copies to conduct experiments or build new plasmids. In other cases, the piece of DNA encodes a useful protein, and the bacteria are used as "factories" to make the protein. For instance, the human insulin gene is expressed in E. coli bacteria to make insulin used by diabetics.

Steps of DNA Cloning

DNA cloning is used for many purposes. As an example, let's see how DNA cloning can be used to synthesize a protein (such as human insulin) in bacteria. The basic steps are:

1. Cut open the plasmid and "paste" in the gene. This process relies on restriction enzymes (which cut DNA) and DNA ligase (which joins DNA).

2. Insert the plasmid into bacteria. Use antibiotic selection to identify the bacteria that took up the plasmid.

3. Grow up lots of plasmid-carrying bacteria and use them as "factories" to make the protein. Harvest the protein from the bacteria and purify it.

Let's take a closer look at each step.

1. Cutting and pasting DNA:

How can pieces of DNA from different sources be joined together? A common method uses two types of enzymes: restriction enzymes and DNA ligase.

A restriction enzyme is a DNA-cutting enzyme that recognizes a specific target sequence and cuts DNA into two pieces at or near that site. Many restriction enzymes produce cut ends with short, single-stranded overhangs. If two molecules have matching overhangs, they can base-pair and stick together. However, they won't combine to form an unbroken DNA molecule until they are joined by DNA ligase, which seals gaps in the DNA backbone.

Our goal in cloning is to insert a target gene (e.g., for human insulin) into a plasmid. Using a carefully chosen restriction enzyme, we digest:

- The plasmid, which has a single cut site.

- The target gene fragment, which has a cut site near each end.

Then, we combine the fragments with DNA ligase, which links them to make a recombinant plasmid containing the gene.

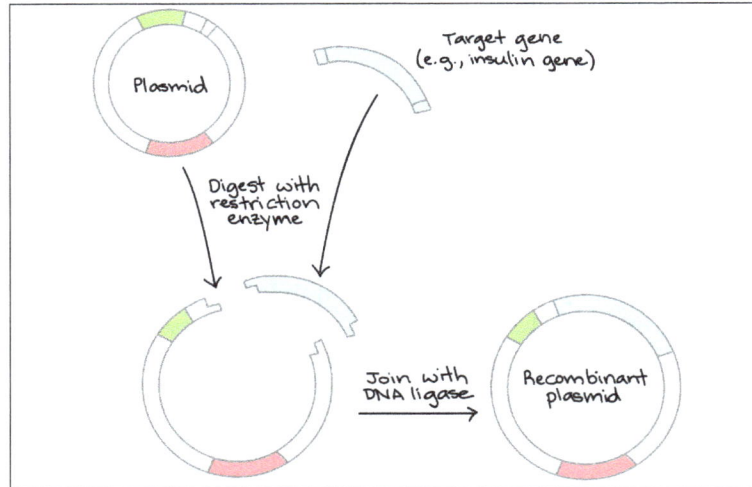

2. Bacterial transformation and selection:

Plasmids and other DNA can be introduced into bacteria, such as the harmless E. coli used in labs, in a process called transformation. During transformation, specially prepared bacterial cells are given a shock (such as high temperature) that encourages them to take up foreign DNA.

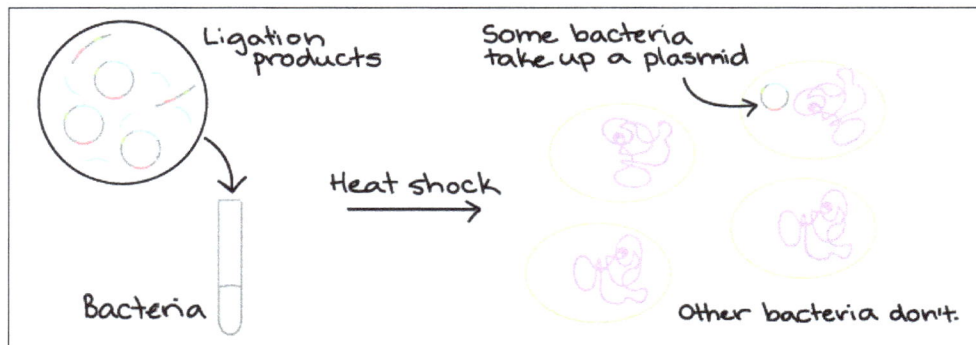

A plasmid typically contains an antibiotic resistance gene, which allows bacteria to survive in the presence of a specific antibiotic. Thus, bacteria that took up the plasmid can be selected on nutrient plates containing the antibiotic. Bacteria without a plasmid will die, while bacteria carrying a plasmid can live and reproduce. Each surviving bacterium will give rise to a small, dot-like group, or colony, of identical bacteria that all carry the same plasmid.

Not all colonies will necessarily contain the right plasmid. That's because, during a ligation, DNA fragments don't always get "pasted" in exactly the way we intend. Instead, we must collect DNA from several colonies and see whether each one contain the right plasmid. Methods like restriction enzyme digestion and PCR are commonly used to check the plasmids.

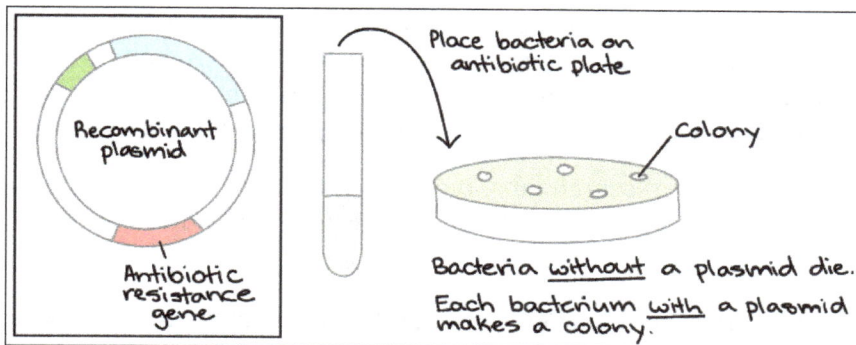

3. Protein production:

Once we have found a bacterial colony with the right plasmid, we can grow a large culture of plasmid-bearing bacteria. Then, we give the bacteria a chemical signal that instructs them to make the target protein.

The bacteria serve as miniature "factories," churning out large amounts of protein. For instance, if our plasmid contained the human insulin gene, the bacteria would start transcribing the gene and translating the mRNA to produce many molecules of human insulin protein.

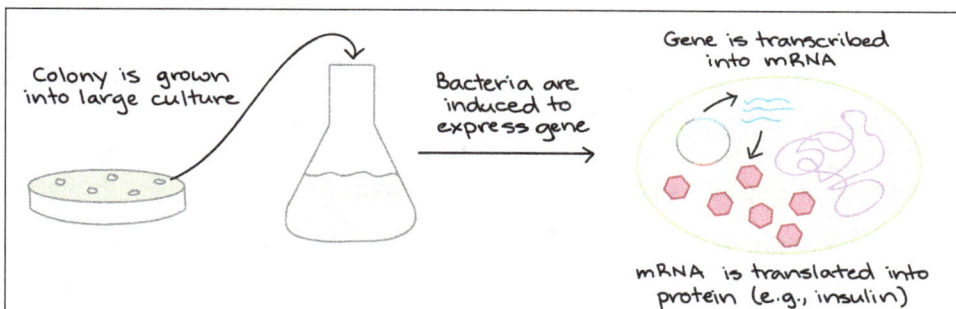

Once the protein has been produced, the bacterial cells can be split open to release it. There are many other proteins and macromolecules floating around in bacteria besides the target protein (e.g., insulin). Because of this, the target protein must be purified, or separated from the other contents of the cells by biochemical techniques. The purified protein can be used for experiments or, in the case of insulin, administered to patients.

Uses of DNA Cloning

DNA molecules built through cloning techniques are used for many purposes in molecular biology. A short list of examples includes:

1. Biopharmaceuticals: DNA cloning can be used to make human proteins with biomedical applications, such as the insulin mentioned above. Other examples of recombinant proteins include human growth hormone, which is given to patients who are unable to synthesize the hormone, and tissue plasminogen activator (tPA), which is used to treat strokes and prevent blood clots. Recombinant proteins like these are often made in bacteria.

2. Gene therapy: In some genetic disorders, patients lack the functional form of a particular

gene. Gene therapy attempts to provide a normal copy of the gene to the cells of a patient's body. For example, DNA cloning was used to build plasmids containing a normal version of the gene that's nonfunctional in cystic fibrosis. When the plasmids were delivered to the lungs of cystic fibrosis patients, lung function deteriorated less quickly.

3. Gene analysis: In basic research labs, biologists often use DNA cloning to build artificial, recombinant versions of genes that help them understand how normal genes in an organism function.

These are just a few examples of how DNA cloning is used in biology today. DNA cloning is a very common technique that is used in a huge variety of molecular biology applications.

Vectors

When geneticists use small pieces of DNA to clone a gene and create a genetically modified organism (GMO), that DNA is called a vector.

What Vectors have to do with Genes and Cloning

In molecular cloning, the vector is a DNA molecule that serves as the carrier for the transfer or insertion of foreign gene(s) into another cell, where it can be replicated and/or expressed. Vectors are among the essential tools for gene cloning and are most useful if they also encode some kind of marker gene encoding a bioindicator molecule that can be measured in a biological assessment to ensure their insertion, and expression, in the host organism.

Specifically, a cloning vector is DNA taken from a virus, plasmid or cells (of higher organisms) to be inserted with a foreign DNA fragment for cloning purposes. Since the cloning vector can be stably maintained in an organism, the vector also contains features that allow for the convenient insertion or removal of DNA. After being cloned into a cloning vector, the DNA fragment can be further sub-cloned into another vector that can be used with even more specificity.

In some cases, viruses are used to infect bacteria. These viruses are called bacteriophages, or phage, for short. Retroviruses are excellent vectors for introducing genes into animal cells. Plasmids, which are circular pieces of DNA, are the most commonly used vectors used to introduce foreign DNA into bacterial cells. They often carry antibiotic resistance genes that can be used to test for expression of the plasmid DNA, on antibiotic Petri plates.

Gene transfer into plant cells is commonly performed using the soil bacterium Agrobacterium tumefaciens, which acts as a vector and inserts a large plasmid into the host cell. Only those cells containing the cloning vector will grow when antibiotics are present.

The classifications are: 1. On the Basis of Our Aim with Gene of Interest 2.On the Basis of Host Cell used 3. On the Basis of Cellular Nature of Host Cell.

On the basis of Our Aim with Gene of Interest

The point is what we are targeting from our gene of interest — its multiple copies or its protein product.

Depending on these criteria vectors are of following two types:

1. Cloning Vectors:

We use a cloning vector when our aim is to just obtain numerous copies (clones) of our gene of interest (hence the name cloning vectors). These are mostly used in construction of gene libraries. A number of organisms can be used as sources for cloning vectors.

Some are created synthetically, as in the case of yeast artificial chromosomes and bacterial artificial chromosomes, while others are taken from bacteria and bacteriophages. In all cases, the vector needs to be genetically modified in order to accommodate the foreign DNA by creating an insertion site where the new DNA will fitted. Example: PUC cloning vectors, pBR322 cloning vectors, etc.

2. Expression Vectors or Expression construct:

We use an expression vector when our aim is to obtain the protein product of our gene of interest. To get the protein we need to allow the expression of our gene of interest (hence the name expression vector) by employing the processes of transcription and translation.

Apart from the three DNA sequences (origin of replication, selectable markers and multiple cloning sites), the expression vectors have some special additional sequences as well. Those are as follows:

   a. A bacterial promoter, such as the lac promoter. The promoter precedes a restriction site where foreign DNA is to be inserted, allowing transcription of foreign sequence to be regulated by adding substances that induce the promoter.

   b. A DNA sequence that, when transcribed into RNA, produces a prokaryotic ribosome binding site.

   c. Prokaryotic transcription initiation and termination sequences.

   d. Sequences that control transcription initiation, such as regulator genes and operators.

In some types of expression vectors which are specifically used in association with the bacterial host (like E. coli), multiple cloning site is not immediately adjacent to the ribosome binding sequence, but instead is preceded by a special sequence coding for a bacterial polypeptide.

While using such type of expression vectors the gene of interest is inserted just after the gene for bacterial polypeptide. In this way we fuse two reading frames, producing a hybrid gene that starts with the bacterial gene and progresses without a break into the codons of our gene of interest.

The product of gene expression is therefore a hybrid protein, consisting of short bacterial polypeptide fused into amino terminus of our target polypeptide sequence. This hybrid polypeptide chain consisting of two different types of polypeptides is called a fusion protein.

The followings are the reasons for incorporation of a fusion protein before our gene of interest:

1. The presence of bacterial peptide at the start of fusion protein may stabilize the molecule and

prevent it from being degraded by the host cell. In contrast the foreign polypeptides that lack a bacterial segment are often destroyed.

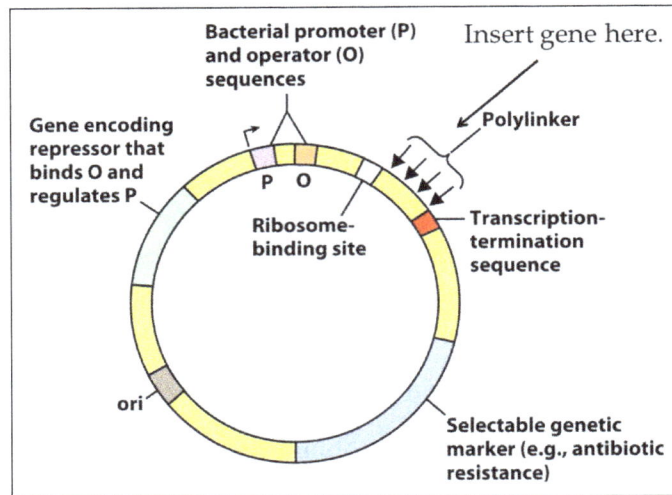

The Key Features of a Typical Expression Vector.

2. The bacterial polypeptide may act as a signal peptide, responsible for transporting our target protein to a specific location from where these are collected. For example, if the bacterial peptides are derived from a protein that is exported by the cell (e.g.; products of ompA genes), then our target polypeptide will simply be transported outside of the host cell straight into the culture media from where these can be collected.

3. The bacterial polypeptide may also help in purification of the target polypeptide by different purification techniques such as affinity chromatography.

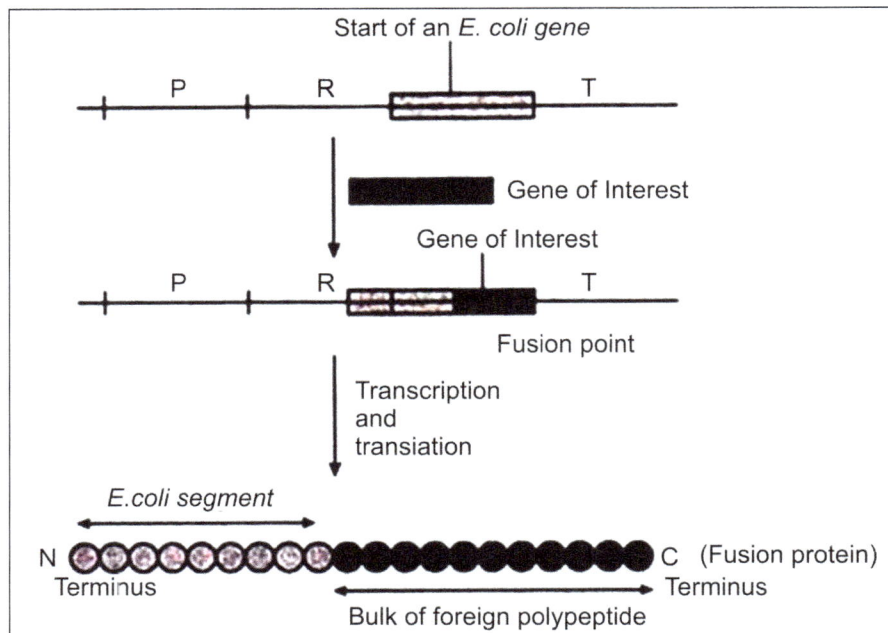

The Construction of a Hybrid Gene and Synthesis of a fusion proteins.

On the basis of Host Cell Used

After construction of a recombinant DNA these can be introduced into a host cell. So depending on the host cell the vectors are designed and constructed. All the parts of the vectors must be functionally compatible with the host. For example, if we are making a vector for a bacterial host, it must have a suitable origin of replication which will be functional in a bacterial cell.

Depending on this Basis the Vectors are Classified as under:

1. Vectors for Bacteria: These are special bacterial origin of replication and antibiotic resistance selectable markers. Bacteria support different kinds of vectors, e.g.; plasmid vectors, bacteriophages vectors, cosmids, phasmids, phagemids, etc.

2. Vectors for Yeast: They have special origin of replication called as autonomously replicating sequences (ARS), e.g., yeast replicative plasmid vectors (YRp) etc.

3. Vectors for Animals: These vectors are needed in biotechnology for the synthesis of recombinant protein from genes that are not expressed correctly when cloned in E. coli or yeast, and methods for cloning in humans are being sought by clinical molecular biologists attempting to devise techniques for gene therapy, in which a disease is treated by introduction of a cloned gene into the patient, e.g., P-element, SV40 etc.

4. Vectors for Plants: The production of genetically modified plants has become possible due to successful use of plant vectors. e.g., Ti-plasmid, Ri-plasmid etc.

On the basis of Cellular Nature of Host Cell

On this basis, the vectors are of two types:

1. Prokaryotic Vectors: This comprises of all vectors for bacterial cells.

2. Eukaryotic Vectors: This comprises of all the vectors for yeast, animal and plant cells.

Prokaryotic Vectors (Bacterial Vectors)

The E. coli cell which is frequently used as a prokaryotic host needs specific types of vectors which are designed accordingly to function in its cytoplasm. Plasmid based and bacteriophage based vectors are most common prokaryotic vectors. The prokaryotic vectors include plasmid derived vectors, bacteriophage derived vectors, phagemid vectors, plasmid vectors and fosmid vectors.

Plasmid Vectors

These are the most common vectors for the prokaryotic host cells. Bacteria are able to express foreign genes inserted into plasmids. Plasmids are small, circular, double- stranded DNA molecules lacking protein coat that naturally exists in the cytoplasm of many strains of bacteria.

Some of the examples of naturally occurring plasmids are Ti plasmids, F-factors, R-factors, Co/E1 plasmid, etc. Plasmids are independent of the chromosome of bacterial cell and range in size from 1000 to 200 000 base pairs. Using the enzymes and 70s ribosomes that the bacterial cell houses, DNA contained in plasmids can be replicated and expressed.

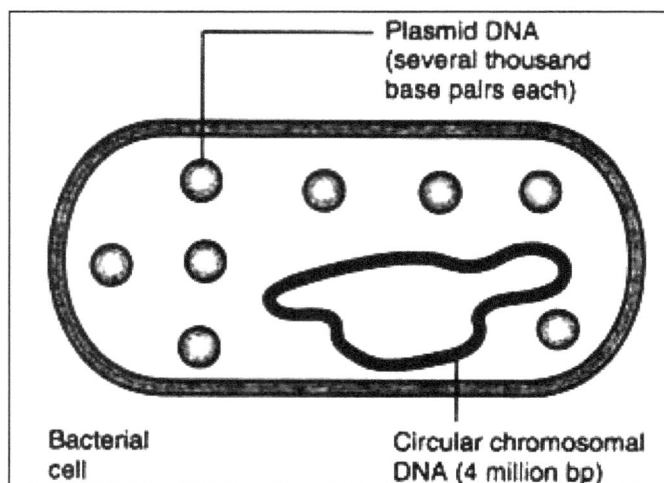

Chromosomal and Plasmid DNA Coexist in an endosymbiotic relationship.

The bacterial cells benefit from the presence of plasmids, which often carry genes that express proteins able to confer antibiotic resistance. These also protect bacteria by carrying genes for resistance to toxic heavy metals, such as mercury, lead, or cadmium.

In addition, some bacteria carry plasmids possessing genes that enable bacteria to break down herbicides, certain industrial chemicals, or the components of petroleum. The relationship between bacteria and plasmids is endosymbiotic; both the bacteria and plasmids benefit from mutual arrangement. Plasmids also possess characteristic copy number.

The higher the copy number, higher is the number of individual plasmids in a host bacterial cell. If more copies of plasmid exist, more protein will be synthesized because of the larger number of gene copies carried by the plasmid. The number of copies plays a role in phenotypic manifestation of a gene. For example, the more copies of an antibiotic-resistance gene there are, the higher the resistance to the antibiotic.

It is very important to note that naturally occurring plasmids do not have all necessary sequences which are required by a DNA molecule to act as a profitable vector. Due to this, natural plasmids are extracted and modified by inserting suitable DNA segments and a complete vector DNA molecule is made.

Plasmid-cloning vectors are derived from bacterial plasmids and are the most widely used, versatile, and easily manipulated ones.

pBR322

This was the first widely used, purpose built plasmid vector. pBR322 has a relatively small size of 4,363 bp. This is important because transformation efficiency is inversely proportional to size and above 10 kbp is very low.

Thus, there is 'room' in pBR322 for an insert of at least six kbp. Also this vector has a reasonably high copy number (~15 copies per cell), which can be increased 200-fold by treatment with a protein-synthesis inhibitor—chloramphenicol amplification.

Nomenclature of pBR 322

The nomenclature of 'pBR 322' can be understood with following explanation:

1. 'p' indicates as a plasmid.

2. 'BR' identifies Bo-liver and Rodriguez, the two researchers who developed it.

3. '322' distinguishes those plasmids from others (like pBR 325, pBR 327, etc.) developed in the same laboratory.

Construction of pBR322

1. Origin of Replication: It carries a fragment of plasmid pMB1 that acts as an origin for DNA replication and thus ensures multiplication of the vector.

2. Selectable Marker: It carries two antibiotic resistance genes—ampicillin and tetracycline.

3. Cloning Sites: It carries a number of unique restriction sites. Some of these are located in one of the antibiotic resistance genes (e.g., sites for Pst I, Pvu I, and Sac I are found in Ampr and BamHI and Hind III in Tetr). Cloning into one of these sites inactivates the gene allowing recombinants to be differentiated from non-recombinants known as insertional inactivation.

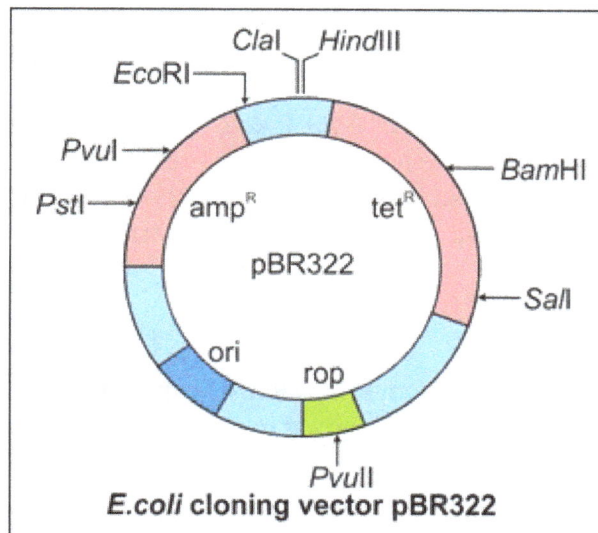

Diagram Sketch of pBR322.

Pedigree of pBR322

By pedigree we understand the origin of pBR322. pBR322 is not a naturally occurring plasmid. It is manufactured by following certain steps which are outlined in figure. It is important to note that pBR322 comprises DNA derived from three different naturally occurring plasmids.

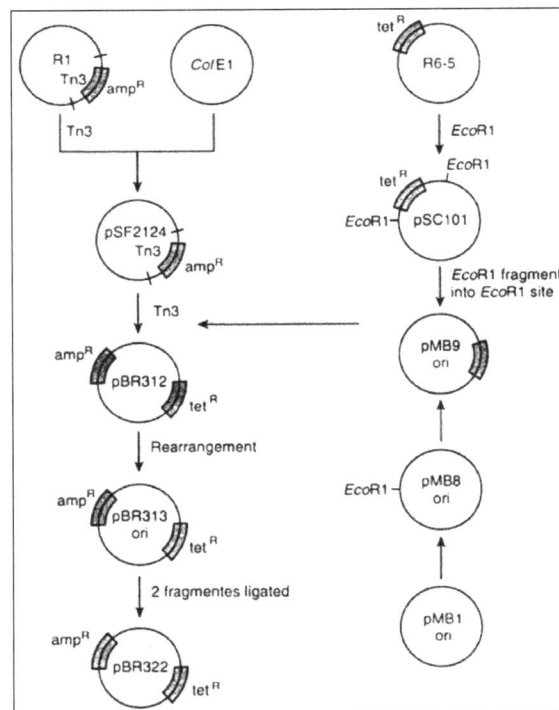

The Pedigree of pBR322.

The amp^R gene originally resided on the plasmid R1 (a naturally occurring antibiotic resistant plasmid in E. coli), the tetR is derived from R6-5 (a second antibiotic resistant plasmid) and the origin of replication is derived from pMB1, which is closely related to the Colicin producing plasmid ColE1.

Recombinant selection with pBR322 – in- sectional inactivation of an antibiotic resistance gene.

When we have introduced our recombinant DNA (vector + gene of interest) into the host cell (by a process called transformation and the host cells that takes up the recombinant DNA are called transformed host cells) then we have to screen the entire host population in order to select the transformed cells (with recombinant DNA) from the non-transformed one (without recombinant DNA).

Every vector has some mechanism associated with it for this screening.

Here we will discuss what is the mechanism followed by the pBR322 vector in this regard. pBR322 has several unique restriction sites that can be used to open up the vector before insertion of a new DNA fragment. BamHl, for example, cuts pBR322 at just one position, within the cluster of genes that code for resistance to tetracycline.

A recombinant pBR322 molecule, one that carries an extra piece of DNA in the BamHl site is no longer able to confer tetracycline resistance on its host, as one of the necessary genes is now disrupted by the inserted DNA. Cells containing this recombinant pBR322 molecule are still resistant to ampicillin, but sensitive to tetracycline ($amp^R\ tef^s$).

Screening for pBR322 recombinants is performed in the following way. After transformation the cells are plated onto ampicillin medium and incubated until colonies appear.

All of these colonies are trans-formants (remember, untransformed cells are amps and so do not produce colonies on the selective medium), but only a few contain recombinant pBR322 molecules: most contain the normal, self-ligated plasmid. To identify the recombinants the colonies are replica plated onto agar medium that contains tetracycline.

After incubation, some of the original colonies regrow, but others do not. Those that do grow consist of cells that carry the normal pBR322 with no inserted DNA and, therefore, a functional tetracycline resistance gene cluster (amp^R tet^R).

The colonies that do not grow on tetracycline agar are recombinants (amp^R tef^s); once their positions are known, samples for further study can be recovered from the original ampicillin agar plate.

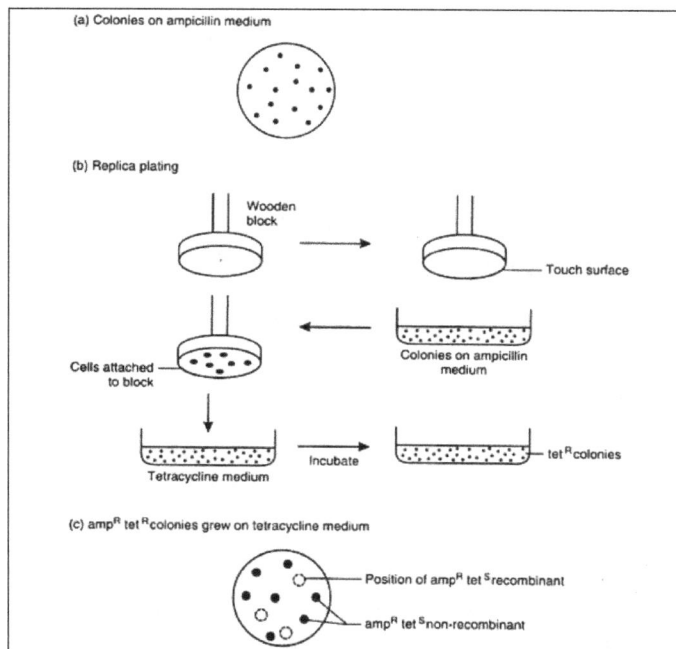

(a) Colonies on ampicillin medium

(b) Replica plating

Wooden block

Touch surface

Cells attached to block

Colonies on ampicillin medium

Tetracycline medium

Incubate

tet^R colonies

(c) amp^R tet^R colonies grew on tetracycline medium

Position of amp^R tet^s recombinant

amp^R tet^s non-recombinant

Screening for pBR322 recombinants by insrtional inactivation of the tetracycline resistance gene.

In figure, (a) Cells are plated onto ampicillin agar: akk tge transformants produce colonies. (b) The colonies are replica plated into tetracycline medium. (c) The colonies that grow on tetracycline medium are (amp^R tet^R) and. Therefore non-recombinant Recombinants (amp^R tef^s) do not grow, but their position on the ampicillin plate is now known.

Uses of pBR322

It is widely used as a cloning vector. In addition to this, it has been widely used as a model system for study of prokaryotic transcription and translation.

Advantages of pBR322

1. Small size ($\sim$ 4.4 kb) enables easy purification and manipulation.

2. Two selectable markers (amp and tet) allow easy selection of recombinant DNA.

3. It can be amplified up to 1000-3000 copies per cell when protein synthesis is blocked by the application of chloramphenicol.

Disadvantages of pBR322

1. It has very high mobility i. e; it can move to another cell in the presence of a conjugative plasmid like F-factor. The nic-bom (bom=basis of mobility) region of pBR322 is responsible for this feature. Due to this, the vector may get lost in a population of mixed host cells.

2. There is a limitation in the size of the gene of interest that it can accommodate.

3. Not a very high copy number is present as is expected from a good vector.

4. Although insertional inactivation of an antibiotic resistance gene provides an effective means of recombinant identification, the method is made inconvenient by the need to carry out two screenings, one with the antibiotic that selects for trans-formants, followed by the second screen, after replica plating, with the antibiotic which distinguishes recombinants.

This makes the screening process time-consuming and laborious.

Another vector pBR327 was derived from pBR322, by deletion of nucleotides between 1,427 to 2,516. These nucleotides are deleted to reduce the size of the vector and to eliminate sequences that were known to interfere with the expression of cloned DNA in eukaryotic cells. pBR327 still contains genes for resistance against two antibiotics (tetracycline and ampicillin).

pBR327 has following two advantages over pBR322:

1. pBR327 has high copy number (30-45 copies per cell).

2. It lacks mobility.

pUC Vectors

pUC are obtained by modifying the pBR322 vector. pUC vectors are smaller than pBR322 of being only ~2.7 kb. But comparatively they have a high copy number. A mutation within the origin of replication produces 500 to 600 copies of the plasmid per cell without amplification.

The nomenclature of 'pUC' can be understood with the following explanation:

1. 'p' indicates the plasmid.

2. 'UC' stands for university of California where it was first developed by J. Messing et al.

We also see many numbers after this like pUC8, pUC18, pUC19 and so on. They are just the series of pUC and have been named just to separate from each other.

The construction of pUC vectors:

1. Origin of Replication: It is derived from the origin of replication of pBR322.The ColE1origin of replication of pBR322 has been modified by carrying out a chance mutation so that each transformed E. coli cell has 500-600 copies of the plasmid.

2. Selectable Marker: It has an ampicillin resistant gene. The transformed host cells can grow on media having ampicillin whereas non-transformed cells die.

3. lac Z' gene having MCS.

4. The lac Z' is incorporated into this vector codes for the enzyme beta-galactosidase which acts on a chromogenic substrate called X-gal (present in bacterial culture media). The expression of lac Z' gene is induced by another compound present in the same media called Iso-propyl-thiogalactoside (IPTG).

When the enzyme substrate reaction takes place, then the X- gal is converted from white to a blue compound. Now the lac Z' gene itself has MCS. Hence, when the gene of interest has been introduced into the lac Z' gene, then it fails to code for beta-galactosidase and thus in this case the substrate (X-gal) is never converted to any other colour. This type of screening is called blue-white screening.

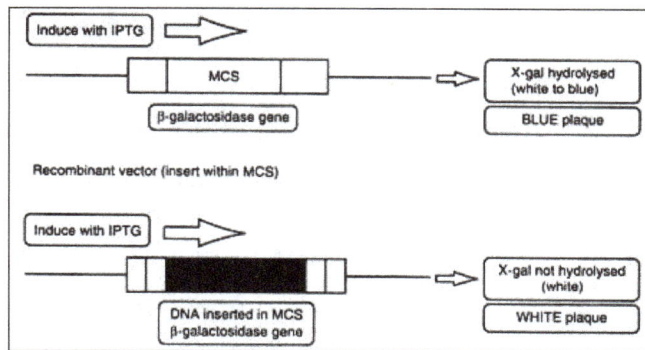

Principle of blue-white selection for the detection of recombinant vectors.

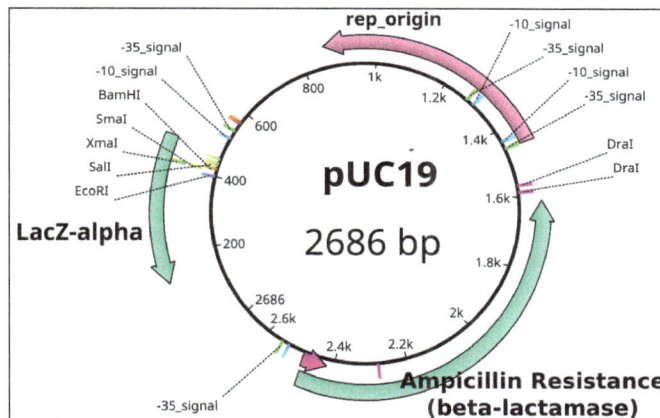

Diagram sketch of pUC18. It differs from the pUC19 only in the region of polylinker site in the arrangement of restriction sites.

Pedigree of pUC Vectors

During the construction of pUC the only selectable marker that is kept out of pBR322 is ampicillin resistant gene. But all the MCS are removed from the amp^R by carrying out chance mutations. The ColE1 origin of replication is also modified by the same process so that it can smoothly carry out the process of replication again and again ultimately increasing the copy number of the vector.

Along with this a lac Z' sequence coding for beta galactosidase is also inserted. Similarly after this by the process of chance mutation we create MCS within the lac Z' sequence.

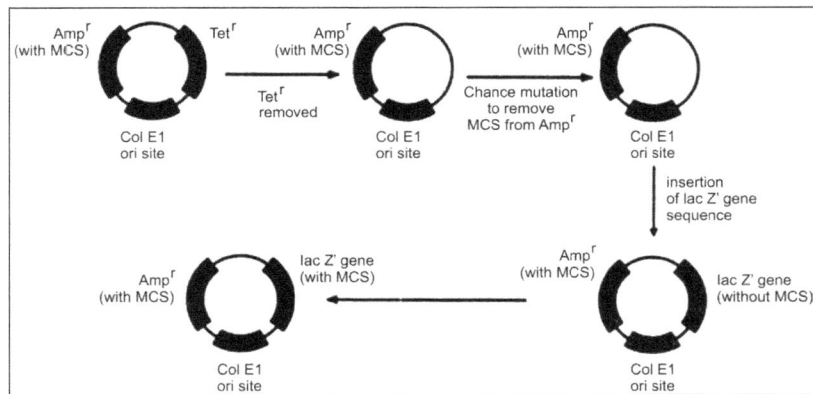

Pedigree of pUC

Screening of Transformed Host Cells using pUC Vectors

After transforming the host cells we carry out their screening to select the transformed cells from non-transformed ones. pUC8, which carries the ampicillin resistance gene and a gene called lac Z', which codes for part of the enzyme beta-galactosidase.

Cloning with pUC8 involves insertional inactivation of the lac Z' gene, with recombinants identified because of their inability to synthesize beta-galactosidase.

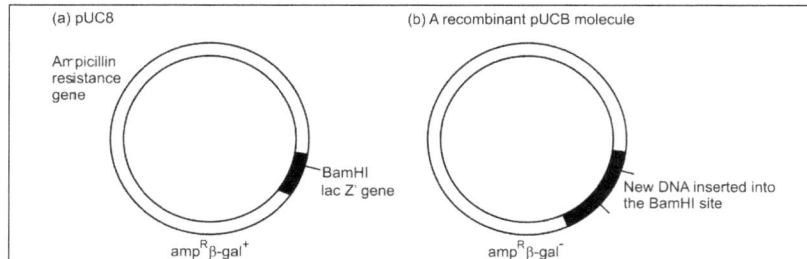

The cloning vectr pUC8:(a)the normal vector molecule; (b) a recombinant pUC8 molecule containing an extra piece of DNA inserted into the BamHI site.

Beta-Galactosidase is one of the series of enzymes involved in breakdown of lactose to glucose plus galactose. It is normally coded by the gene lac Z, which resides on E. coli chromosome. Some strains of E. coli have a modified lac Z gene, one that lacks the segment referred to as lac Z' and coding for the a-peptide portion of beta-galactosidase.

These mutants can synthesize the enzyme only when they harbour a plasmid, such as pUC8, that carries the missing lac Z' segment of the gene.

A cloning experiment with pUC8 involves selection of trans-formants on ampicillin agar followed by screening for beta-galactosidase activity to identify recombinants. Cells that harbour a normal pUC plasmid are ampR and able to synthesize beta-galactosidase; recombinants are also ampR but unable to make beta-galactosidase.

Screening for beta-galactosidase presence or absence is, in fact, quite easy. Rather than assay for lactose being split to glucose and galactose, we test for a slightly different reaction that is also catalysed by beta-galactosidase.

This involves a lactose analogue called X-gal (5-bromo-4-chloro-3-indolyl-beta-D- galactopyrano-side) which is broken down by beta-galactosidase to a product that is coloured deep blue.

If X-gal (plus an inducer of the enzyme such as Iso-pro-pylthiogalactoside, IPTG) is added to the agar, along with ampicillin, then non-recombinant colonies, the cells of which synthesize beta-ga-lactosidase, will be coloured blue, whereas recombinants with a disrupted lac Z' gene and unable to make p-galactosidase, will be white.

This system, which is called Lac selection, is summarized in. Note that both ampicillin resistance and the presence or absence of p-galactosidase is tested on a single agar plate. The two screenings are, therefore, carried out together and there is no need for the time-consuming replica-plating step that is necessary with plasmids such as pBR322.

Some commercially available plasmid vectors			
		Applications	
pBR322	AP′ Tc′ Single cloning sites	General cloning and subclon-ing in E. Coli	Various
pAR153	AP′ Tc′ Single cloning sites	General cloning and subclon-ing in E. Coli	Various
pGEM®.3Z	AP′ Tc′ MCS SP6/T7 promoters lac Za – peptide	General cloning and in vitro trancscription in E. Coli and single-stranded DNA produc-tion	Promega
pCRl⊖	AP′ MCS T7 promoters CMV enhancer/promoter	Expression of genes in mam-malian cells	Promega
pET.3	AP′ MCS T7 promoters	Expression of genes in bacte-rial cells	stratagene
pCMV – Script⊖	Neo′ MCS CMV enhancer/promoter	High level expression of genes in mammalian cells and clon-ing of PCR products	stratagene

Uses of pUC Vectors

pUC vectors can be used both as cloning vector and expression vector. When used as an expression vector its sequences are slightly modified to meet necessary requirements.

Advantages of pUC Vectors

The pUC vectors offer following major advantages over pBR322 vectors:

- High copy number of 500-600 copies per cell.

- Easy and single step selection.

- The unique restriction sites used for cloning are clustered within the MCS. This allows cloning of a DNA fragment having two different sticky ends.

Disadvantages of pUC

It cannot accommodate a gene of interest larger than 15kb.

Fosmid Vectors

These are similar to cosmids but are based on the bacterial F-plasmid. The cloning vector is limited, as a host (usually E. coli) can only contain one fosmid molecule. Fosmids are 40 kb of random genomic DNA. Fosmid library is prepared from a genome of target organism and cloned into a fosmid vector.

Low copy number offers higher stability than comparable high copy number cosmids. Fosmid system may be useful for constructing stable libraries from complex genomes.

Bacteriophage Derived Vectors

Bacteriophages, or phages as they are commonly known, are viruses that specifically infect bacteria. Like all viruses, phages are very simple in structure, consisting merely of a DNA (or occasionally ribonucleic acid (RNA)) molecule carrying a number of genes, including several for replication of the phage, surrounded by protective coat or capsid made up of protein molecules.

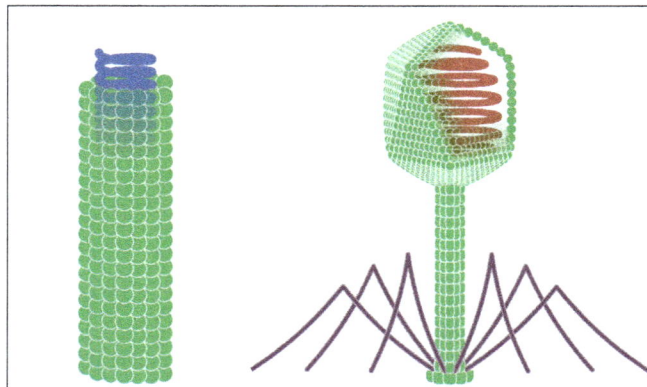

Two main types of phage structure: (a) head-and-tail; (b) filamentous.

The general pattern of infection, which is the same for all types of phage, is a three-step process:

- The phage particle attaches to the outside of bacterium and injects its DNA chromosome into the cell.

- The phage DNA molecule is replicated, usually by specific phage enzymes coded by genes on the phage chromosome.

- Other phage genes direct synthesis of protein components of capsid, and new phage particles are assembled and released from the bacterium.

With some phage types the entire infection cycle is completed very quickly, possibly in less than 20min. This type of rapid infection is called a lytic cycle, as release of the new phage particles is associated with lysis of the bacterial cell.

The characteristic feature of a lytic infection cycle is that phage DNA replication is immediately followed by synthesis of capsid proteins, and the phage DNA molecule is never maintained in a stable condition in the host cell. In contrast to a lytic cycle, lysogenic infection is characterized by retention of the phage DNA molecule in the host bacterium, possibly for many thousands of cell divisions.

Fred Blatter and his colleagues were the first to develop a bacteriophage as vector.

Bacteriophage M13 vectors

General Biology

The M13 family of vectors is derived from bacteriophage M13. This is a male specific (infects E. coli having f. pili), lysogenic filamentous phage with a circular single-stranded DNA genome about 6,407 bp (6.4 kb) in length. Once inside the host-cell the single-stranded DNA of M13 phage acts as the template for synthesis of a complementary strand, resulting in normal double-stranded DNA.

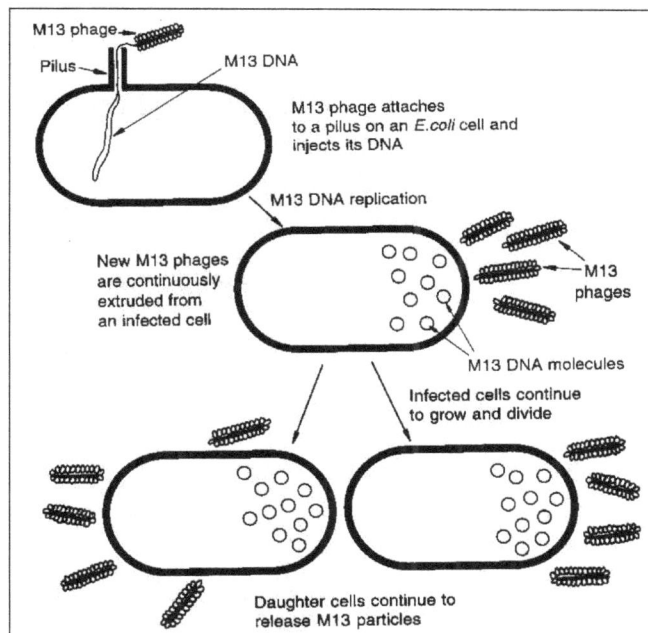

The Life Cycle of M13 Bacteriophage.

This molecule is not inserted into the bacterial genome, but instead replicates until over 100 copies are present in the cell. When the bacterium divides, each daughter cell receives copies of the phage genome, which continues to replicate, thereby maintaining its overall numbers per cell.

As shown in figure, new phage particles are continuously assembled and released, about 1000 new phages being produced during each generation of an infected cell.

Attraction of M13 as a Cloning Vector

Several features of M13 make this phage attractive as the basis for a cloning vector. The genome is less than 10 kb in size, well within the range desirable for a potential vector. In addition, the double-stranded replicative form (RF) of the M13 genome behaves very much like a plasmid, and can be treated as such for experimental purposes.

It is easily prepared from a culture of infected E. coli cells and can be reintroduced by transfection. Most importantly, genes cloned with an M13-based vector can be obtained in the form of single-stranded DNA. Single-stranded version of cloned genes are useful for several techniques, notably DNA sequencing and in vitro mutagenesis.

Using an M13 vector is an easy and reliable way of obtaining single-stranded DNA for this type of work.

Construction of M13 Vectors

The first step in the construction of M13 cloning vector is to introduce the lac Z' gene into the inter-genic sequence. This gives rise to M13 mp1 which forms blue plaques on X-gal agar M13 mp1 does not progress any unit restriction site in the lac Z' gene.

It however, contains a hexanucleotide sequence GGATTC near the start of the gene. A single nucleotide change (by using in vitro mutagenesis) would make this GAATTC, which is an EcoR1 site.

This results in the formation of M13 mp2. M13 mp2 has a slightly altered lac Z' gene but the beta-galactosidase enzyme produced by cells infected with M13 mp2 is still perfectly functional. M13 mp2 is the simplest M13 vector. DNA fragments with ECoR1 sticky ends can be inserted into the cloning site and recombinants are distinguished as clear plaques on X-gal agar.

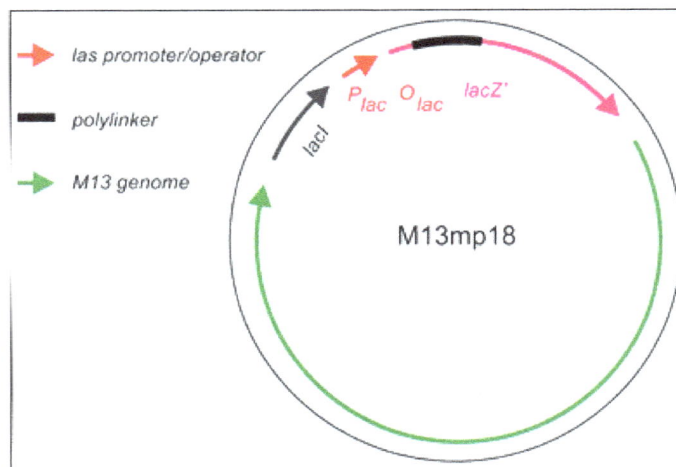

M13 Vector.

We go for further modifications of M13 mp2 resulting in the production of another M13 vector called M13 mp7. In the generation of M13 mp7 first of all we synthesize a short oligonucleotide called poly-linker that consists of a series of restriction sites and has EcoR1 sticky ends. This poly-linker is inserted into the EcoRI site of M13 mp2 to generate M13mp7.

This poly-linker also provides as many as four possible cloning sites (ECoRI, BamHl, SaiI and PstI) to the new vector. It is very important to note that the poly-linker is designed so that it does not totally disrupt the lac Z' gene: a reading frame is maintained throughout the poly-linker, and a functional, though altered, beta-galactosidase enzyme is still produced.

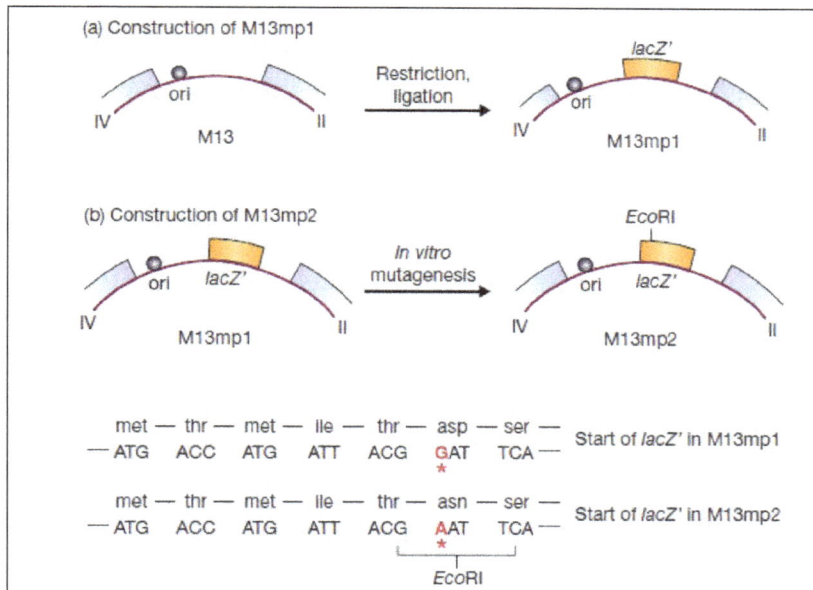

The Construction of M13mp2.

Screening of transformed host cells using bacteriophage M13 vectors:

Insertion of new DNA almost invariably prevents beta-galactosidase production. So recombinant plaques are clear on X-gal agar. Alternatively, if the poly-linker is reinserted, and the original M13mp7 reformed, then blue plaques result.

Uses of Bacteriophage M13 Vectors

1. DNA Sequencing:

For a long time the most important application of M13 cloning was in DNA sequence determination by the Sanger method, also called the dideoxy or chain-termination method. This relies on synthesis of DNA in the presence of chain terminating inhibitors, the 2′, 3′- di-deoxynucleoside triphosphates (ddNTPs). The method is now a very standard tool of molecular biology.

2. Phage Display Vectors:

An important use of filamentous phage is in phage display systems. Here, coding sequences are inserted into one of the coat protein genes. The result is that the phage are generated with a hybrid form of this protein, which is a fusion of the normal protein sequence and the protein product of the inserted sequence (assuming the inserted sequence has the same reading frame as the coat protein gene).

The phages are secreted from the cell, with this extra material 'displayed' on the outside. These display vectors have many uses, e.g., in screening libraries by panning and for vaccine production.

3. Other Applications:

Some protocols for site-directed mutagenesis also use single- stranded DNA, which can be obtained with vectors based on filamentous phages. Single-stranded DNA is also of particular use in generating probes for RNA analysis.

Probes can be prepared that are specific for RNA transcripts from either strand of DNA. The latter applications are outside the scope of this book, but more information can be obtained from specialized laboratory manuals.

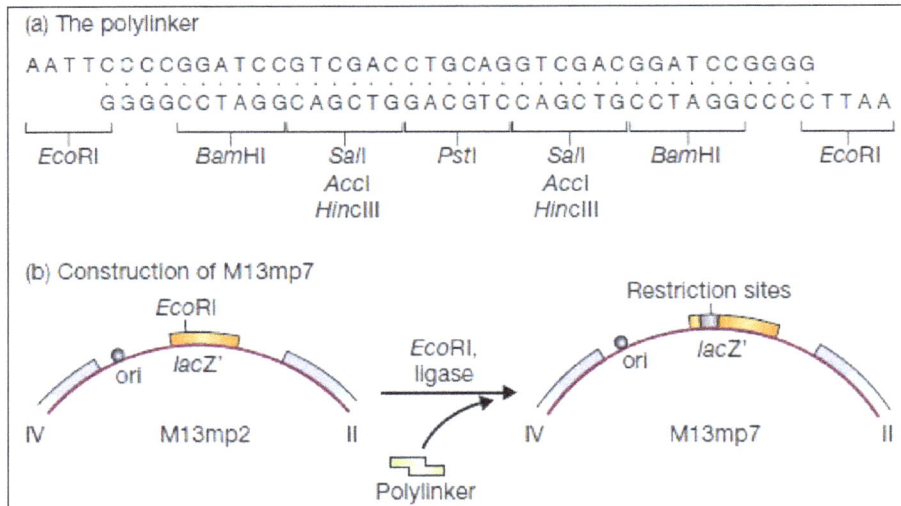

The Construction of M13mp7 from M13mp2.

Advantages of Bacteriophage M13 Vector

1. Advantages over Lambda Phage Vector:

M13 is an example of a filamentous phage and is completely different in structure from lambda. Furthermore, the M13 DNA molecule is much smaller than the lambda genome, being only 6407 nucleotides in length. It is circular and is unusual in that it consists entirely of single- stranded DNA.

The smaller size of the M13 DNA molecule means that it has room for fewer genes than the lambda genome. This is possible because the M13 capsid is constructed from multiple copies of just three proteins (requiring only three genes), whereas synthesis of the lambda head- and-tail structure involves over 15 different proteins.

In addition, M13 follows a simpler infection cycle than lambda , and does not need genes for insertion into the host genome. Injection of an M13 DNA molecule into an E. coli cell occurs via the pilus, the structure that connects two cells during sexual conjugation.

2. Other Advantages:

M13-based vectors are that they contain the same poly-linker and alpha-peptide fragments as the pUC series and recombinants can be selected by the blue → white colour test. Also the size of the genome is below 10 kb and so is easy to handle.

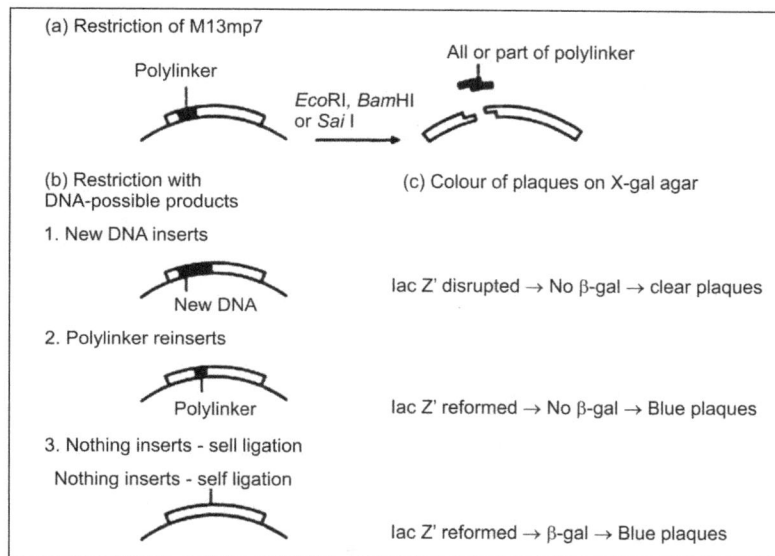

Cloning with M13mp7.

Disadvantages of Bacteriophage M13 Vectors

The following are the disadvantages of bacteriophage M13 vectors:

- Gene of interest more than 2kb cannot be cloned.

- It has low yield of DNA.

- The phage produce many toxins in high concentration.

Lamda Phage Vectors

This is a widely used vector for the cloning of very large pieces of genes.

General Biology

Lambda is a typical example of a head-and-tail phage. The genetic material is DNA which is present in the polyhedral head structure and the tail serves to attach the phage to the bacterial surface and to inject the DNA into the cell. The lambda DNA molecule is 49 kb in size.

It is a temperate phase and this can carry out lytic and lysogenic cycles. The positions and identities of most of the genes on the lambda DNA molecule are known.

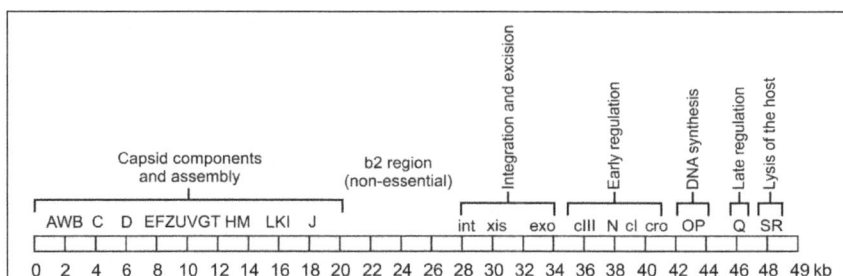

The Genetic Map of Lambda DNA.

Lambda phage can have both linear and circular forms of DNA. The molecule shown in is linear, with two free ends, and represents the DNA present in the phage head. This linear molecule consists of two complementary strands of DNA, base paired according to the Watson-Crick rules.' However, at either end of the molecule is a short 12-nucleotide stretch in which the DNA is single-stranded.

The two single strands are complementary, and so can base pair with one another to form a circular, completely double-stranded molecule. Complementary single strands are often referred to as 'sticky' ends or cohesive ends, because base pairing between them can 'stick' together the two ends of a DNA molecule (or the ends of two different DNA molecules).

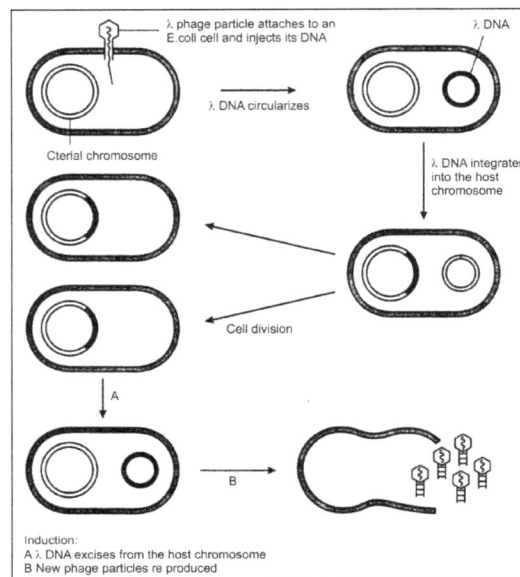

Lysogenic Cycle of Lambda Phage.

The lambda cohesive ends are called the cos sites and they play two distinct roles during the lambda infection cycle. First, they allow the linear DNA molecule that is injected into the cell to be circularized, which is a necessary prerequisite for insertion into the bacterial genome.

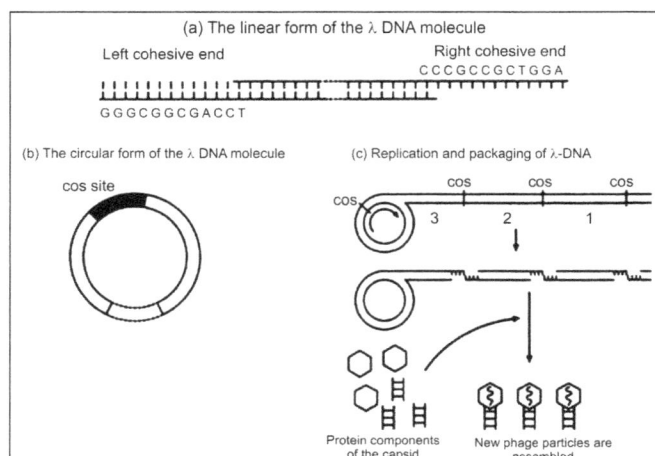

Linear and Circular Forms of Lambda DNA . (a)The linear form ,showing the left and right cohesive ends. (b) Base pairing between the cohesive ends results in the circular form of the molecule, which are individually packaged into lambda phage heads as new lambde particies are assembled.

The second role of the cos sites is rather different, and comes into play after the pro-phage has excised from the host genome. At this stage a large number of new lambda DNA molecules are produced by the rolling circle mechanism of replication, in which a continuous DNA strand is rolled off the template molecule. The result is a catenane consisting of a series of linear A genomes joined together at the cos sites.

The role of the cos sites is now to act as recognition sequences for an endonuclease that cleaves the catenane at the cos sites, producing individual lambda genomes. This endonuclease, which is the product of gene A on the lambda DNA molecule, creates the single stranded sticky ends, and also acts in association with other proteins to package each lambda genome into a phage head structure.

The cleavage and packaging processes recognize just the cos sites and the DNA sequences to either side of them. Changing the structure of the internal regions of the lambda genome, for example, by inserting new genes has no effect on these events so long as the overall length of the A genome is not altered too greatly.

Problems associated with naturally occurring lambda phage to be used as cloning vectors:

Two problems have to be solved before lambda- based cloning vectors could be developed:

1. The lambda molecule can be increased in size by only about 5%, representing the addition of only 3kb of new DNA. If the total size of the molecule is more than 52kb, then it cannot be packaged into the lambda head structure and infective phage particles are not formed. This severely limits the size of a DNA fragment that can be inserted into an unmodified lambda vector.

2. The lambda genome is so large that it has more than one recognition sequence for virtually every restriction endonucleases. Restriction cannot be used to cleave the normal lambda molecule in a way that will allow insertion of new DNA, because the molecules would be cut into several small fragments that would be very unlikely to reform a viable lambda genome on relegation.

Due to these reasons the DNA of naturally occurring lambda phage cannot be used as a cloning vector. To solve this issue we modify the lambda's genome and make it suitable to be a successful vector.

Solving the Problems

1. From research it has been found out that large segment in the central region of the lambda DNA molecule can be removed without affecting the ability of the phage to infect E. coli cells. Removal of this nonessential region between positions 20 and 35 on the map decreases the size of the lambda genome by up to 15kb.

This makes a room for as much as 18kb of new DNA which can be added to it to form a recombinant molecule.

This non-essential genes thus removed are involved in integration and excision of the lambda prophage from the E. coli chromosome. A deleted lambda genome is, therefore, non-lysogenic and can follow only the lytic infection cycle. This in itself is desirable for a cloning vector as it means that we can get plaques (a visible structure formed within a cell culture, such as bacterial cultures within some nutrient medium).

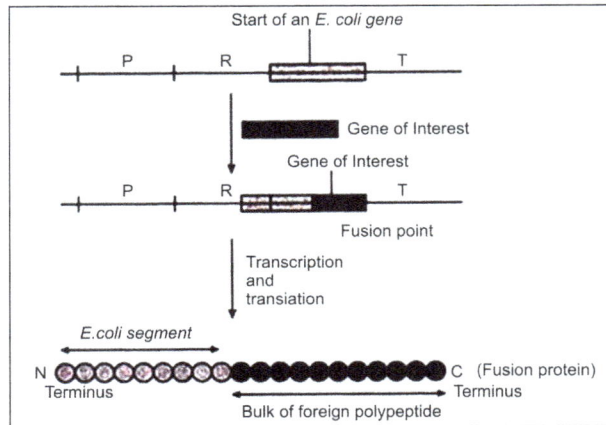

Solutions of Two Problems Associated with using Lambda Genome as Cloning Vector.

2. We can remove unnecessary restriction sites by carrying out in vitro mutagenesis. For example, an ECoRI site, GAATTC, could be changed to GGATTC, which is not recognized by the enzyme.

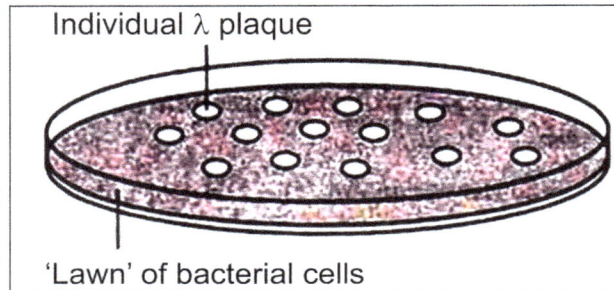

The plaques formation due to the lytic cycle aried out by lambda phage.

Types of Lambda Vectors

There are two types of lambda cloning vectors.

1.　Lambda Insertion Vectors:

In this case a large segment of the non-essential region has been deleted, and the two arms ligated together. An insertion vector possesses at least one unique restriction site into which new DNA can be inserted. The size of the DNA fragment that an individual vector can carry depends on the extent to which the non-essential region has been deleted, e.g.; lambda-gtlo, lambda- ZAP11.

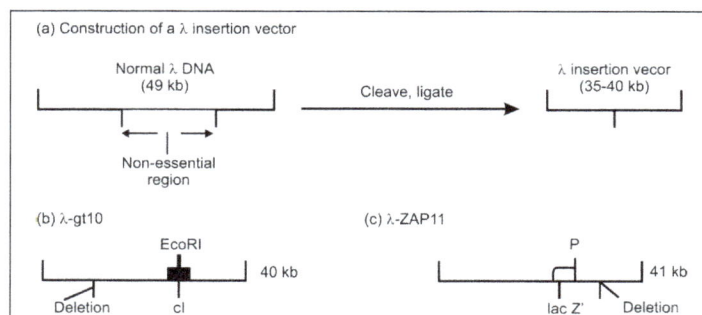

Lamcda insertion vectors: P - polylinker in the lac Z gene of lambda ZAP ll,
containing unique restriction site for Sacl. Notl , Xbal, etc.

2. Lambda Replacement Vectors:

These vectors have two recognition sites for the restriction endonucleases. These sites flank a segment of DNA that is replaced by the DNA to be cloned. Often the replaceable fragment (or stuffer fragment) carries additional restriction sites that can be used to cut it up into small pieces so that its own reinsertion during a cloning experiment is very unlikely.

Replacement vectors are generally designed to carry large pieces of DNA than insertion vectors can handle e.g., lambda- EMBL, lambda-GEMll, etc.

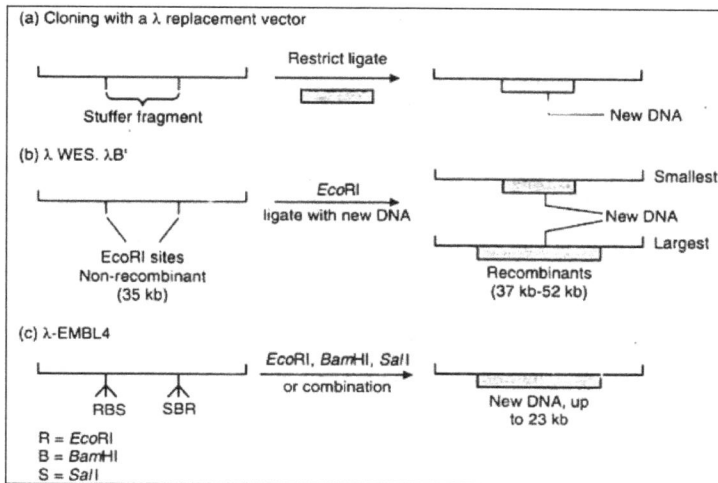

Lambda Replacement Vectors.

Cloning Experiments with Lambda Insertion or Replacement Vectors

A cloning experiment with a lambda vector can be carried out by following the similar method that we followed for a plasmid vector—the lambda DNA molecules are digested with suitable restriction endonuclease enzyme, the gene of interest is added, the mixture is ligated and the resulting recombinant DNA is introduced into E. coli host cell (by a process called transfection).

This type of experiment requires that the vector be in its circular form, with the cos sites hydrogen bonded to each other.

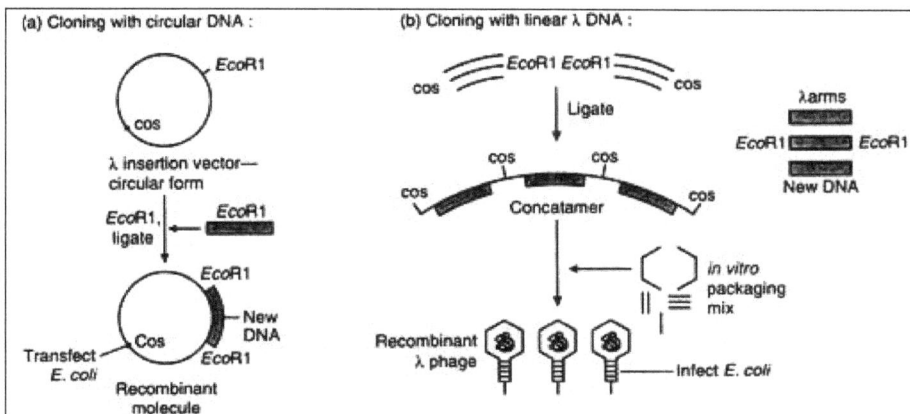

Different Strategies for Cloning with Lambda Vector.

The transfection process which requires a circular lambda DNA molecule is not particularly efficient. To obtain a greater number of recombinants we can introduce some refinements in the lambda genome. In this regard we can prefer a linear form of the vector. When the linear form of the vector is digested with the relevant restriction endonuclease, the left and right arms are released as separate fragments.

A recombinant DNA can be constructed by mixing together our gene of interest with the vector arms. Ligation results in several molecular arrangements, including catenae's comprising left arm-DNA-right arm repeated many times. Recombinant phage thus produced in the test tube can be used to infect an E.coli culture.

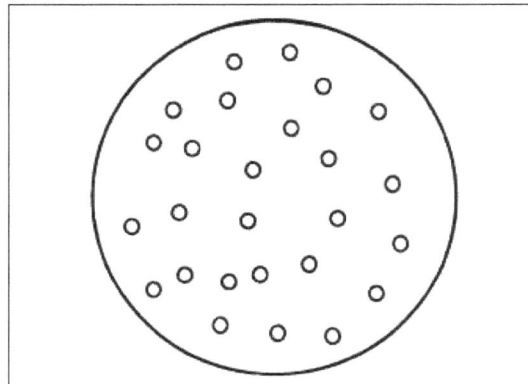

Bacteriphage Plaques.

Visualization of Phage Infection after the Process of Transfection

The entry of recombinant DNA in the host cell is followed by the lytic cycle which eventually results in the lysis of the host cell. The lysed host cell can be located on the agar medium as plaques on a lawn of bacteria. Each plaque is a zone of clearing produced as the phages lyse the cell and move on to infect the neighbouring bacteria.

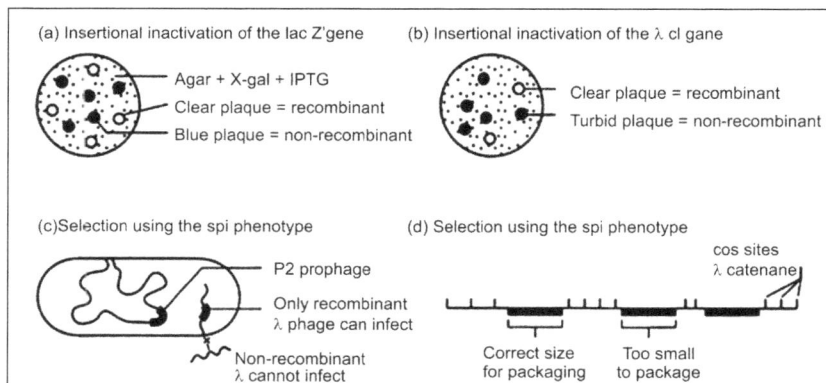

(a) Insertional inactivation of the lac Z'gene
 - Agar + X-gal + IPTG
 - Clear plaque = recombinant
 - Blue plaque = non-recombinant

(b) Insertional inactivation of the λ cI gane
 - Clear plaque = recombinant
 - Turbid plaque = non-recombinant

(c) Selection using the spi phenotype
 - P2 prophage
 - Only recombinant λ phage can infect
 - Non-recombinant λ cannot infect

(d) Selection using the spi phenotype
 - cos sites
 - λ catenane
 - Correct size for packaging
 - Too small to package

Strategies for the Selection of Recombinant Plaques Using Lambda Vector.

Screening of Transformed Host Cells using Bacteriophage Lambda Vectors

A variety of ways could be employed to distinguish between recombinant plaques from non- recombinant ones.

The methods are as follows:

1. Insertional Inactivation of Lac Z' Gene Carried by the Lambda Phage Vector:

Insertion of our gene of interest into the lac Z' gene inactivates beta- galactosidase synthesis. Recombinants are distinguished by plating cells on X-gal agar where the recombinant plaques are clear whereas non-recombinant plaques are blue in colour.

2. Insertional Inactivation of Lambda Cl Gene:

Several lambda cloning vectors have restriction site in the cl gene. Insertional inactivation of cl gene cause a visible change in the plaque morphology. Normal plaques appear turbid (hazy) whereas recombinant plaques with disrupted cl gene are clear.

3. Selection using Spi Phenotype:

P2 phage is a relative of lambda phage, lamda phages cannot infect E. coli cells that already has an integrated P2 phage in its genome. Due to this, lamda phage is said to be Spi−+ (sensitive to P2 prophage infection). Some lambda cloning vectors are designed so that insertion of new DNA causes a change from Spi−+ to Spi−−, enabling the recombinant to infect cells that carry P2 pro-phages.

Such cells are used as host for cloning experiments with these vectors. In this case recombinants are Spi-, so they are able to form plaques.

4. Selection on the Basis of Lambda Genome Size:

We know this from the beginning that any gene of interest which is less than 37kb or more than 52kb cannot be packed in the head of lambda phage. Many lambda vectors have been constructed by deleting large segments of the lambda DNA molecule and so are less than 37kb in length.

These can only be packaged into mature phage particles after our gene of interest has been inserted. This brings the total genome size up to 37kb or more. Hence, with these vectors only recombinant phages are able to replicate.

Uses of Bacteriophage Lambda Vectors

The main use of all lambda based vectors is to clone DNA fragments that are too long to be handled by plasmid or M13 vectors. A replacement vector such as lambda-EMBL4 can carry up to 20kb of our gene of interest. This compares with a maximum insert size of about 8kb for almost all plasmids and less than 3kb for M13 vectors.

Advantages of Bacteriophage Lambda Vectors

Following are the advantages of lambda vectors:

1. Storage of phage particles is comparatively much easier than that of plasmid based vectors.

2. The shelf-life of phage particles is infinite.

3. Transfection of E. coli is much easier with phage particles.

Disadvantages of Bacteriophage Lambda Vectors

If you have isolated a clone, it is frequently quite difficult to isolate large quantities of DNA. In practice, many problems are encountered that do not occur with plasmids. There is still no truly rapid, reliable protocol for the production of very clean lambda-DNA.

The most successful method is to use anion exchange columns. The most frequent problem is that the preparation contains dirt that makes further processing, such as a restriction digestion, difficult or impossible.

Even in the replacement vectors, almost two thirds of the DNA is made up of vector sequences. If possible, you should clone the sections that are of interest using plasmids. LambdaZAP banks can save work, because the plasmid portions are cut out in vivo, along with inserted DNA. That process is highly efficient, requires only a relatively few work steps, and lasts only 1 to 2 days.

Some commercially available bacteriophage –based vectors			
λGTll	λ insertion vector insert capacity 7.2 Kbp lac Z gene	cDNA library cronsturction Expression of inserts	Various
λEMVBL 3/4	λ insertion vector insert capacity9-23kbp	Genomic libracy onstruction	Various
λZAPExoress[⊖]	λ-based insertion vector capacity of 12 kbp in vivo excision of inserts Expression of inserts	cDNA library cronsturction also genomic/PCR cloning	Strata-gene
λFIX[⊖]ll	λ-based replacement vector: capacity 9-23 kbp Spi./ P2 seledtion system to reduce ono-re-combinant background	Genomic library construction	Strata-gene
pBluescript[⊖]l l	Phagemid vector produces single-stranded DNA	I vitro transcription DNA sequencing	Strata-gene
Super Cos l	Cosmid vector with Ap′ and Neo′ markers, plus T3 and T7 promoters Capacity 30-42 kbps	Generation of cosmid-Based ge-nomic DNA libraries T3/7 pro-moters allow end-specific chro-mosome walking techniques	Strata-gene

Cosmid Vectors

It is the most sophisticated type of lambda based vector. Cosmids are the hybrids between the phage DNA molecule and bacterial plasmid. Their design centres on the fact that the enzymes that package the lambda DNA molecule into the phage protein coat need only the cos sites in order to function.

Construction of Cosmid Vectors

A cosmid is basically a plasmid that carries a cos site. It also needs a selectable marker, such as ampicillin resistant gene, and a plasmid origin of replication. This is important to note that as cosmid lacks all the lambda genes, so at does not produce plaques. Instead colonies are formed on the selective media just as with plasmid vectors.

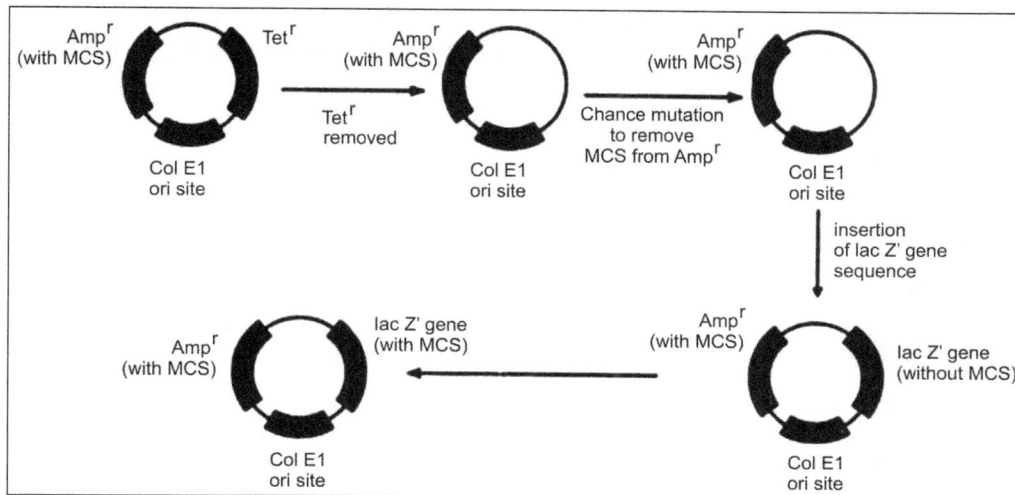

A typical Cosmid vector and way it is used to clone long fragment of DNA.

Cloning Experiment with Cosmid Vectors

This is carried out as follows. The cosmid is opened and its unique restriction site and our gene of interest is inserted. These fragments are usually produced by partial digestion with a restriction endonuclease, as total digestion almost invariably results in fragments that are too small to be cloned with a cosmid.

Ligation is carried out so that catenanes are formed. These lambda phages are then used to infect an E. coli culture. All colonies are recombinant colonies as non-recombinant lambda phages cannot be packaged into the head of the lambda bacteriophage.

Uses of Cosmid Vectors

Cosmids are used for construction of genomic libraries of eukaryotes since these can be used for cloning large fragments of DNA.

Advantages of Cosmids

Followings are advantages of cosmid vectors:

1. These can be used to clone gene of interest up to 40 kb.

2. As the lambda phage will insert the recombinant DNA into the host cell, an extra step of inserting the recombinant DNA into the host cell is not performed.

3. Easy screening method is found.

Phagemid Vectors

Although M13 vectors are very useful for the production of single-stranded versions of cloned genes they do have one disadvantage. There is a limit to the size of DNA fragment that can be cloned with an M13 vector, with 1500bp generally being looked on as the maximum capacity.

To get around this problem a number of novel vectors have been constructed which are the hybrids of plasmids and M13 vectors. We call them phagemids ('phage' from M13 bacteriophage and 'mid' from plasmid).

Construction of Phagemid Vector

A typical phagemid has following parts:

1. Phage M13 origin of replication.

2. A portion of lac Z' gene driven by lac promoter.

3. A multiple cloning site (MCS) with lac Z' gene.

4. Phage T7 and T3 promoter sequences flanking the MCS sequences.

5. ColE1 origin of replication.

6. amp^R resistant gene.

Plasmids that carry the M13 replication origin in addition to a conventional origin of dsDNA synthesis can be replicated either as dsDNA from the latter or as single-stranded DNA from the M13 origin. Replication from the M13 origin requires the appropriate proteins (such as gene II protein) to be provided from a helper phage also replicating within the cell.

Replication generates single-stranded DNA which can then be packaged into phage coats. Examples of phagemids are the vectors pUC118, 119 and 120.

They are replicated as plasmids until the cell containing them is co-infected with a helper phage, such as M13K07, which provides the proteins for single-stranded DNA synthesis and packaging. M13K07 is an M13 phage that has been modified, most importantly by the incorporation of a plasmid replication origin.

Replication from this origin allows the helper phage to be present in a high copy number per cell and, therefore, to provide the larger quantities of the proteins that are required to replicate and package the phagemid molecule.

M13K07 also contains a kanamycin resistance gene to allow for selection for the presence of the helper phage. (Of course, it is possible that the M13K07 helper phage may be packaged too, but in practice the packaged phagemid molecules are found to be in a 100-fold excess over the helper phage.)

Another example of these vectors is the pBluescript series, such as pBluescriptIIKSÞ, shown in figure. This series of plasmids contains, in addition to features already described, promoters from the E. coli bacteriophages T3 or T7, which are useful for expressing cloned sequences.

The phagemid contains an ampicillin resistande salectale marker (εmpR) and origins (ori) for double-stranded DNA synthesis(Phage), the latter for use when cells containing the vector areco-infected with a suitable helper phage. There are a multiple cloning site (MCS)in the lacZ ,omogene (allowing blue-white selection for the presence of an insert) and phage T3 and T7 promoters for transcription of inserted DNA sequences.

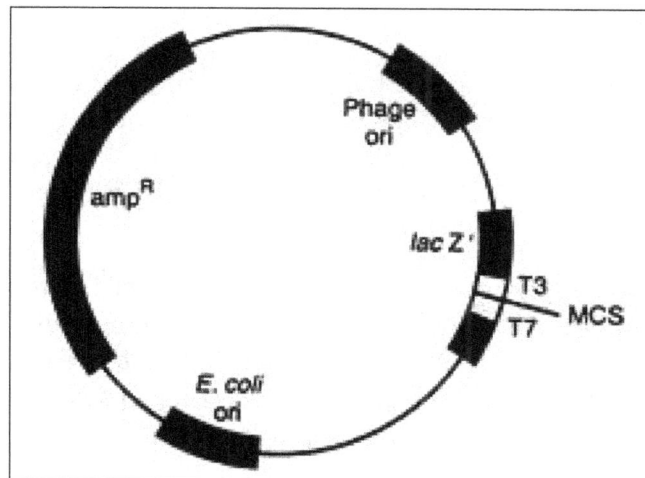

Uses of Phagemid Vectors

This vector is a multipurpose vector as it can serve as following:

1. A cloning vector

2. An expression vector

3. A sequencing vector

Advantages of Phagemid Vectors

The main advantage of the phagemid system is that it can be used to provide single-or double-stranded material without any re-cloning.

Phasmid Vectors

Phasmids are truly plasmids with phage genes. These are linear duplex DNAs whose ends are lambda segments that contain all the genes required for a lytic infection and whose middle-portion is linearized. Both the lambda and the plasmid replication functions are intact.

Normally, plasmid vectors carry a lambda attachment site. Once inside E. coli cell, the phasmid can replicate like a phage and form plaques in the normal way. However, if the vector contains the gene that encodes lambda repressor, then the plasmid replicates as a plasmid rather than as a phage.

Depending upon the functioning or non-functioning of cl- Protein (coded by repressor), the phasmid can replicate as plasmid (cl-Protein inactive) or phage when cl-protein is active. The activity of cl-protein can be inactive by growing the E.coli culture at 40 °C.

Plasmids may be used in variety of ways. For example, DNA may be cloned in the plasmid vector in a conventional way and then the recombinant plasmid can be lifted onto the phage. Plasmids are easy to store, they have an effectively infinite shelf life and screening phages by molecular hybridization gives cleaner results than screening bacterial colonies.

Eukaryotic Vectors

Most cloning experiments are carried out with E. coli as the host, and the widest variety of cloning vectors are available for this organism. E. coli is particularly popular when the aim of the cloning experiment is to study the basic features of molecular biology such as gene structure and function. However, under some circumstances it may be desirable to use a different host for a gene cloning experiment.

But when the aim of the RDT experiment is not just to study a gene but to use cloning to control or improve synthesis of an important metabolic product (e.g., a hormone such as insulin), or to change the properties of the organism (e.g., to introduce herbicide resistance into a crop plant), then we take a host cell which is more advanced and capable of meeting an advanced level of metabolism.

Due to this, many times we consider eukaryotic vectors for our cloning experiments. Yeast, animal and plant vectors are all considered as eukaryotic vectors.

Different types of cloning vectors and the size of the inserted DNA that can be accommodated	
Types of vector	Insert DNA size (in kb)
Plasmid	0.5 to 8
Bacteriophage Lambda	9 to 25
Cosmid	30 to 45
YAC	250 to 1000
BAC	50 to 300

Yeast Vectors

The yeast (Saccharomyces cerevisiae) is one of the most important organisms in biotechnology. Its role is also very important in brewing and bread making. Yeast has been used as a host organism for the production of important pharmaceuticals from cloned genes. Development of cloning vectors for yeast has been stimulated greatly by the discovery of plasmid that is present in most strains of S. cerevisiae.

General Construction

Various yeast vectors have been designed, once the ability and utility of yeast is confirmed. All of them have three features in common.

1. All of them contain unique target sites for a number of restriction endonucleases.

2. All of them can replicate in E. coli often at high copy number.

3. All of them employ markers that can be used to select recombinant yeasts, e.g., Hi53, leo2, trpl and ura3.

Types of Yeast Vectors

All the yeast vectors can be divided into three types:

1. Yeast cloning vectors (or Yeast plasmid vectors).

2. Yeast expression vectors.

3. Yeast artificial chromosomes (YAC).

Yeast Cloning Vectors (or Yeast Plasmid Vectors)

These vectors are used to clone (make several duplicate copies) our gene of interest in the yeast host cell. All the cloning vectors have been engineered from 2fx plasmid which is the naturally occurring plasmid in the yeast cell.

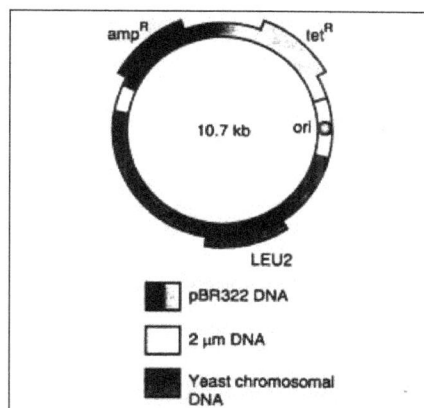

Structure of YEps

They are of following types:

1. Yeast Episomal Plasmids (YEps): It is 6,318 bp long and has a copy number of 70-200. Most of the YEps are shuttle vectors (can be used as vectors both in prokaryotic hosts and eukaryotic hosts) and thus have been engineered accordingly. An example of YEps is Yep13.

It have following parts:

- Origin of replication derived from 2m plasmid.

- ampR and tetR region from pBR322. This selectable marker region is helpful for the screening of recombinants host cells when the vector is used in a prokaryotic cell.

- LEU2 region which is derived from yeast chromosome and could be used as a selectable when the vector is used in an eukaryotic cell. LEU2, which codes for beta-iso-propyl-malate dehydrogenase, one of the enzymes involved in the conversion of pyruvic acid to leucine.

This is very important to note that when we are taking yeast cells for YEps then we have to take only leu2 yeasts which must be auxotrophic mutants having non-functional LEU2 gene.

Screening Process:

The leu2- yeast host cells are unable to synthesize amino acid leucine and can survive, only if this amino acid is supplied as a nutrient in the growth medium. Selection is possible because trans-formants host cells contain a copy of YEp (having LEU2 gene) and are quite able to grow in the absence of amino acid.

In a cloning experiment, cells Eire plated out onto minimal medium, which contains no added amino acids. Only transformed cells are able to survive and form colonies.

YEp may get inserted into the yeast chromosome by a process of homologous recombination between the plasmid LEU2 gene and the yeast mutant LEU2 gene. The plasmid may remain integrated, or a later recombination event may result in it being excised again.

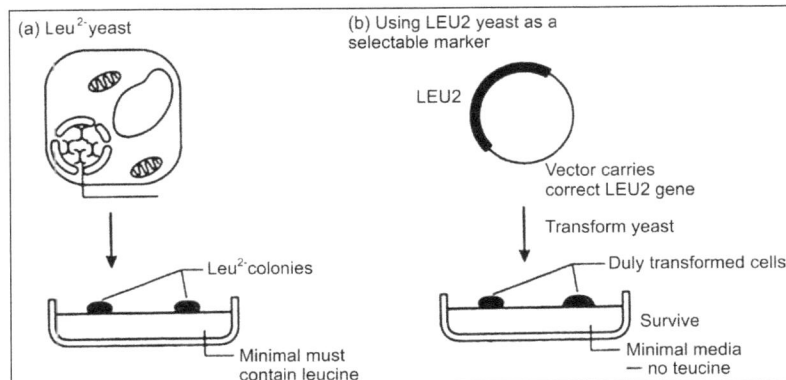

Using the LEU2 gene as a selectable marker in a yeast cloning esperiment.

2. Yeast Integrative Plasmids (YIps): These are basically bacterial plasmids carrying a yeast gene. An example is YIp5, which is pBR322 with an inserted URA3 gene.

3. Yeast Replicative Plasmids (YRps): These are able to multiply as independent plasmids because they carry a chromosomal DNA sequence that includes an origin of replication. An example is YRp7.

Yeast Expression Vectors

These vectors are used when our aim is to express our gene of interest in the yeast cell. Yeast expression vectors will employ promoter and terminator sequences in addition to the gene of interest. Apart from these we have genetic tags like the gene for green fluorescent protein (GFP) for tracking the location of the protein after its biosynthesis.

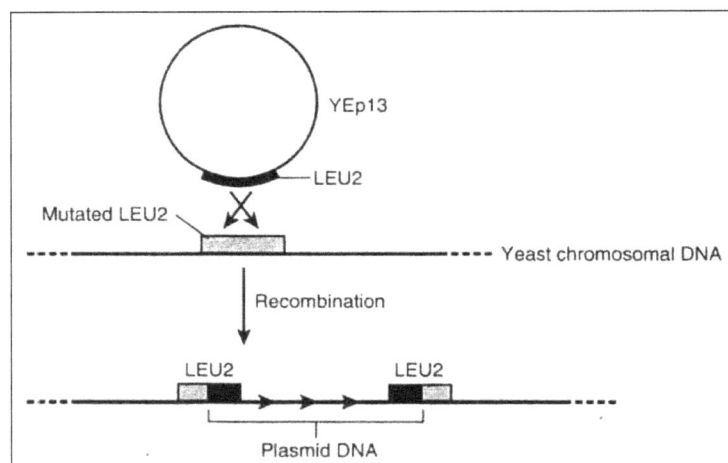

Recombination between plasmid and chromosomal LEU2 genes can integrate Yep13 into yeast chromosomal DNA.

Following are few examples:

1. p427-TEF: High copy yeast expression vector carrying the aminoglycoside phosphotransferase gene for selection in yeast using G418. Inserts are expressed from the strong TEF promoter.

2. p417-CYC: Low copy yeast expression vector carrying the aminoglycoside phosphotransferase gene for selection in yeast using G418. Inserts are expressed from the weak CYC1 promoter.

3. PTEF-MF: Yeast expression vector for secreted proteins. A strong TEF1 promoter drives constitutive expression of a cDNA fused to the pre-pro leader sequence of mating factor alpha to ensure secretion of the protein product into the medium.

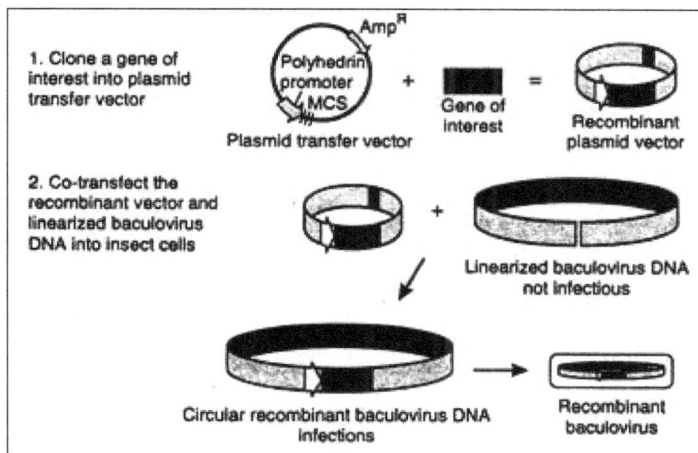

Generation of Recombinant Baculovirus.

Yeast Artificial Chromosomes

Yeast artificial chromosomes (YACs) are synthetic double-stranded linear constructs containing the elements necessary for replication as independent chromosomes in yeast.

Animal Vectors

The Followings are Some Examples of Animal Vectors:

1. Baculovirus Vector:

Baculovirus infects insects. This virus is rod shaped with a large double-stranded genome. During normal infections, baculovirus produces nuclear inclusion bodies which consist of virus particles embedded in a protein matrix.

This protein matrix is called polyhedrin and it accounts for 70% of total protein encoded by the virus. Genetic manipulation of the viral DNA is not possible as it has a very large DNA with many restriction sites for a single enzyme.

Hence, the gene of interest is cloned into the small recombination transfer vector and co-transfected into insect cell lines along with the wild type of virus in the cell. Homologous recombination takes place between the polyhedrin gene and our gene of interest.

Thus, our gene of interest will be transferred from the vector plasmid into the wild type of virus and polyhedrin gene will be transferred from the virus on the plasmid.

This is something like displacement reaction. This displacement of gene will not affect the replication of virus, as polyhedrin gene is not required for replication. The recombination virus replicates in the cells and generates characteristic plaques (without inclusion bodies).

Normally the virus is cultured in the insect cell line of Spodopterafrugiperda. The gene of interest is expressed during the infection and very high yields of protein can be achieved by the time the cell lyses.

2. Bovine Papilloma Virus Vector:

Bovine papilloma virus (BPV) causes warts (uncontrolled epithelial proliferation) in cattle.

BPV normally infects terminally differen tiated squamous epithelial cells. BPV has a capsid protein surrounding a circular double-stranded DNA of size 79 kb. 69% of this genome is important for viral function, whereas 31 % of the genome can be replaced by any foreign DNA sequence like our gene of interest.

The recombinant BPV is constructed by ligating our gene of interest and BPV vector (69%) onto the pBR 322 plasmid, thus generating the shuttle vector containing plasmid ori site and virus replication sequences. These shuttle vectors are multiplied in E. coli cells first and then they are transformed into mouse cell line.

The major advantage of BPV is the generation of permanent cell line. As the infected cells are not killed, a stable plasmid number is found even when the insert is of large size. The selection of transform-ants is very easy as they form a pile of cells on the transferred monolayer of cells called "Focus".

The transformed cells are then selected by the presence of marker gene which is mostly the neomycin phospho transferase gene coding for resistance against G418. Example of this type of vector is p3.7LDL.

3. SV40 Virus based Vectors:

SV 40 is a spherical virus with double-stranded circular DNA of size 5.2 kb. The viral protein contains three viral coded proteins. VP1 is the major protein present in the capsid with a size of 47000 kDa. Two more proteins VP2 and VP3 are also present.

The DNA of virus is associated with the four histones (H4, H2A, H2B and H3) proteins. The viral DNA can be segmented into five precise segments coding for five different proteins small T, large T, VP1, VP2 and VP3. VP1 coding region overlaps VP2 and VP3 in a different translation reading frame. SV 40 virus infects monkey kidney cell lines.

The virus travels to the nucleus and gets uncoated. Then both the T-genes located near the origin are translated in the clockwise direction. The large T protein is important for virus DNA replication and starts after the translation of large T -protein.

Replication starts at the origin and is bi-directional. It terminates when two replication forks meet. About 105 molecules of duplex DNA are synthesized per cell. Along with DNA replication, VP1, VP2 and VP3 proteins are synthesized.

Then packing of DNA occurs to form new virions, which are released by the lysis of cell. The entire process can also be initiated by transfection with naked SV 40 DNA. SV 40 vectors are constructed similar to phage vectors. Portions of the viral genome are removed and replaced by other DNA segments. There are three types of SV 40 vehicles each of which have a distinct advantage or disadvantage among themselves.

They are as follows:

(a) SV40 Passive Transforming Vectors:

These vectors neither replicate nor produce virions, but simply integrate the DNA segments into the cellular DNA. These transformed cells replicate the new DNA as an integral part of their own genomes. These plasmids are also shuttle vectors and include selective markers like herpes virus, thymidine kinase or neo genes.

Apart from the selective markers, they include transcriptional regulator signals and polyadenylation sites.

(b) SV40 Trans-ducting Vectors:

These vectors are capable of replicating and packing into virion particles. Transducing vectors contain a segment of 300 bp which functions as the origin of replication and provides the transcriptional regulatory signals for the synthesis of mRNAs.

This type of vector takes an insert of size 3.9 to 4.5 kb. These plasmids do not have the genes that code for VP1, VP2 and VP3. As no DNA can be added to SV 40 DNA without removing any DNA from the genome, to add the insert, the genome DNA that is not required is removed.

The functions of the DNA that are lost by these deletions are supplied by using a helper virus or by inserting the SV 40 deleted genes into the host DNA. Normally the recombinant SV 40 vectors (usually consist of DNA of interest and replication sequence and gene for coding VP1, VP2 and VP3) are transformed into the cos cell line.

Cos cell line is a kidney cell line of the African green monkey kidney. It has the T-protein gene incorporated in the genome. So when the vector is transfected into these cells, virion particles are yielded with the help of helper virus.

(c) SV40 Plasmid Vectors:

These vectors multiply in the monkey cell line but are not packed as the virions. These vectors usually contain origin of replication sequences and larger T-protein gene but do not contain VP1, VP2 and VP3 genes. They are shuttle vectors, and have the ability to multiply both in E. coli and monkey cell line.

Plant Vectors

Cloning vectors for higher plants were developed in 1980s and their use has led to the genetically modified (GM) crops that are in the headline today.

We will examine the details of plant vectors and the genetic modification of crops. Here we look at the cloning vectors and how they are used.

Three types of cloning system have been used with varying degrees of success with higher plants:

1. Vectors based on naturally occurring plasmids of Agrobacterium (e.g., Ti plasmids from A. tumifaciens and Ri plasmid from A. rhizogens).

2. Direct gene transfer using various types of plasmid DNA. (e.g., using of supercoiled plasmids).

3. Vectors based on plant viruses (e.g., Caulimo virus vectors and Gemini virus vectors).

Shuttle Vectors

Shuttle vectors are those which can multiply into two different unrelated species. Shuttle reactors are designed to replicate in the cells of two species, as they contain two origins of replication, one appropriate for each species as well as genes that are required for replication and not supplied by the host cell, i.e., it is self-sufficient with the process of its replication.

DNA Sequencing

DNA sequencing is the process of determining the sequence of nucleotide bases (As, Ts, Cs, and Gs) in a piece of DNA. Today, with the right equipment and materials, sequencing a short piece of DNA is relatively straightforward.

Sequencing an entire genome (all of an organism's DNA) remains a complex task. It requires breaking the DNA of the genome into many smaller pieces, sequencing the pieces, and assembling the sequences into a single long "consensus." However, thanks to new methods that have been developed over the past two decades, genome sequencing is now much faster and less expensive than it was during the Human Genome Project.

We'll take a look at methods used for DNA sequencing. We'll focus on one well-established method, Sanger sequencing, but we'll also discuss new ("next-generation") methods that have reduced the cost and accelerated the speed of large-scale sequencing.

Sanger Sequencing: The Chain Termination Method

Regions of DNA up to about 900 base pairs in length are routinely sequenced using a method called Sanger sequencing or the chain termination method. Sanger sequencing was developed by the British biochemist Fred Sanger and his colleagues.

In the Human Genome Project, Sanger sequencing was used to determine the sequences of many relatively small fragments of human DNA. (These fragments weren't necessarily 900 bp or less, but researchers were able to "walk" along each fragment using multiple rounds of Sanger sequencing.) The fragments were aligned based on overlapping portions to assemble the sequences of larger regions of DNA and, eventually, entire chromosomes.

Although genomes are now typically sequenced using other methods that are faster and less expensive,

Sanger sequencing is still in wide use for the sequencing of individual pieces of DNA, such as fragments used in DNA cloning or generated through polymerase chain reaction (PCR).

Ingredients for Sanger Sequencing

Sanger sequencing involves making many copies of a target DNA region. Its ingredients are similar to those needed for DNA replication in an organism, or for polymerase chain reaction (PCR), which copies DNA in vitro. They include:

- A DNA polymerase enzyme.

- A primer, which is a short piece of single-stranded DNA that binds to the template DNA and acts as a "starter" for the polymerase.

- The four DNA nucleotides (dATP, dTTP, dCTP, dGTP).

- The template DNA to be sequenced.

However, a Sanger sequencing reaction also contains a unique ingredient:

- Dideoxy, or chain-terminating, versions of all four nucleotides (ddATP, ddTTP, ddCTP, ddGTP), each labeled with a different color of dye.

- Dideoxy nucleotides are similar to regular, or deoxy, nucleotides, but with one key difference: they lack a hydroxyl group on the 3' carbon of the sugar ring. In a regular nucleotide, the 3' hydroxyl group acts as a "hook," allowing a new nucleotide to be added to an existing chain.

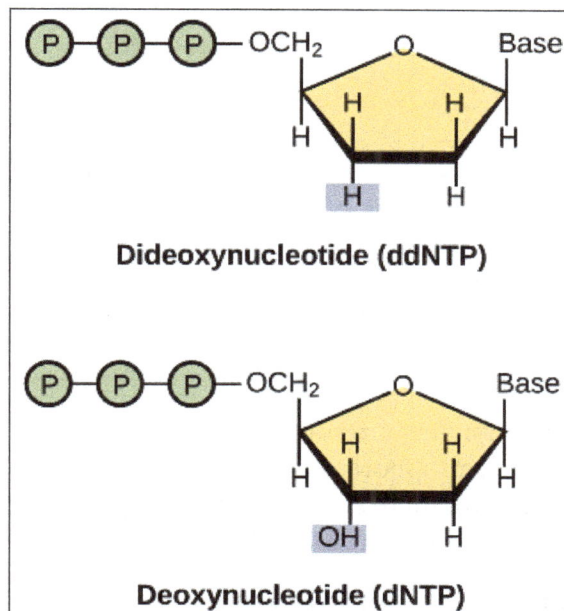

Whole-genome sequencing

Once a dideoxy nucleotide has been added to the chain, there is no hydroxyl available and no further nucleotides can be added. The chain ends with the dideoxy nucleotide, which is marked with a particular color of dye depending on the base (A, T, C or G) that it carries.

Method of Sanger sequencing

The DNA sample to be sequenced is combined in a tube with primer, DNA polymerase, and DNA nucleotides (dATP, dTTP, dGTP, and dCTP). The four dye-labeled, chain-terminating dideoxy nucleotides are added as well, but in much smaller amounts than the ordinary nucleotides.

The mixture is first heated to denature the template DNA (separate the strands), then cooled so that the primer can bind to the single-stranded template. Once the primer has bound, the temperature is raised again, allowing DNA polymerase to synthesize new DNA starting from the primer. DNA polymerase will continue adding nucleotides to the chain until it happens to add a dideoxy nucleotide instead of a normal one. At that point, no further nucleotides can be added, so the strand will end with the dideoxy nucleotide.

This process is repeated in a number of cycles. By the time the cycling is complete, it's virtually guaranteed that a dideoxy nucleotide will have been incorporated at every single position of the target DNA in at least one reaction. That is, the tube will contain fragments of different lengths, ending at each of the nucleotide positions in the original DNA. The ends of the fragments will be labeled with dyes that indicate their final nucleotide.

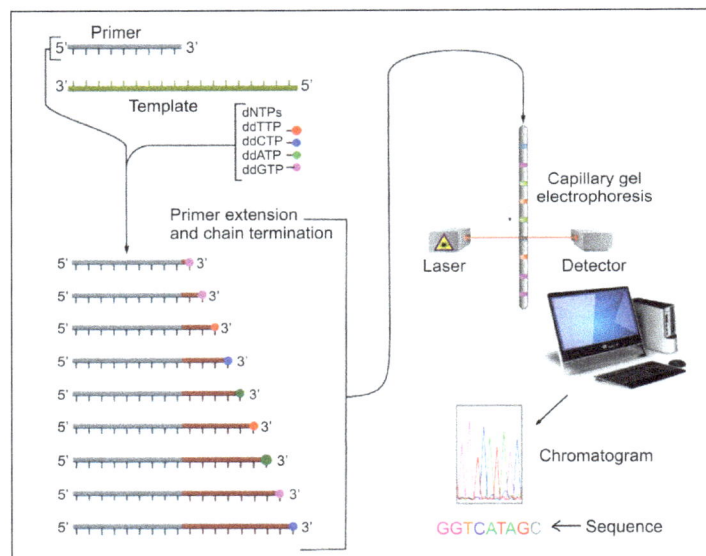

After the reaction is done, the fragments are run through a long, thin tube containing a gel matrix in a process called capillary gel electrophoresis. Short fragments move quickly through the pores of the gel, while long fragments move more slowly. As each fragment crosses the "finish line" at the end of the tube, it's illuminated by a laser, allowing the attached dye to be detected.

The smallest fragment (ending just one nucleotide after the primer) crosses the finish line first, followed by the next-smallest fragment (ending two nucleotides after the primer), and so forth. Thus, from the colors of dyes registered one after another on the detector, the sequence of the original piece of DNA can be built up one nucleotide at a time. The data recorded by the detector consist of a series of peaks in fluorescence intensity, as shown in the chromatogram above. The DNA sequence is read from the peaks in the chromatogram.

Uses and Limitations

Sanger sequencing gives high-quality sequence for relatively long stretches of DNA (up to about 900 base pairs). It's typically used to sequence individual pieces of DNA, such as bacterial plasmids or DNA copied in PCR.

However, Sanger sequencing is expensive and inefficient for larger-scale projects, such as the sequencing of an entire genome or metagenome (the "collective genome" of a microbial community). For tasks such as these, new, large-scale sequencing techniques are faster and less expensive.

Next-generation Sequencing

There are a variety of next-generation sequencing techniques that use different technologies. However, most share a common set of features that distinguish them from Sanger sequencing:

- Highly parallel: Many sequencing reactions take place at the same time.

- Micro scale: Reactions are tiny and many can be done at once on a chip.

- Fast: Because reactions are done in parallel, results are ready much faster.

- Low-cost: Sequencing a genome is cheaper than with Sanger sequencing.

- Shorter length: Reads typically range from 50-700 nucleotides in length.

Conceptually, next-generation sequencing is kind of like running a very large number of tiny Sanger sequencing reactions in parallel. Thanks to this parallelization and small scale, large quantities of DNA can be sequenced much more quickly and cheaply with next-generation methods than with Sanger sequencing. For example, in 2001, the cost of sequencing a human genome was almost $100 million. In 2015, it was just $1245.

Why does fast and inexpensive sequencing matter? The ability to routinely sequence genomes opens new possibilities for biology research and biomedical applications. For example, low-cost sequencing is a step towards personalized medicine – that is, medical treatment tailored to an individual's needs, based on the gene variants in his or her genome.

DNA Replication

There are about 3 billion base pairs of DNA in your genome, all of which must be accurately copied when any one of your trillions of cells divides. The basic mechanisms of DNA replication are similar across organisms.

DNA replication is semiconservative, meaning that each strand in the DNA double helix acts as a template for the synthesis of a new, complementary strand.

This process takes us from one starting molecule to two "daughter" molecules, with each newly formed double helix containing one new and one old strand.

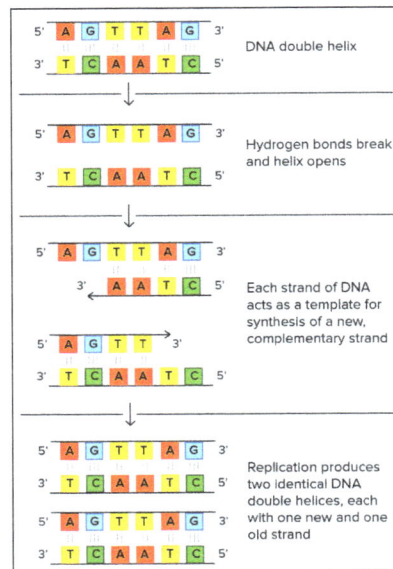

In a sense, that's all there is to DNA replication. But what's actually most interesting about this process is how it's carried out in a cell.

Cells need to copy their DNA very quickly, and with very few errors (or risk problem such as cancer). To do so, they use a variety of enzymes and proteins, which work together to make sure DNA replication is performed smoothly and accurately.

DNA Polymerase

One of the key molecules in DNA replication is the enzyme DNA polymerase. DNA polymerases are responsible for synthesizing DNA: they add nucleotides one by one to the growing DNA chain, incorporating only those that are complementary to the template.

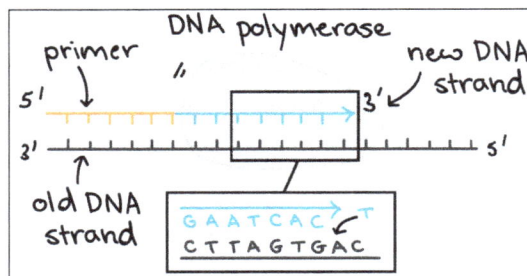

Here are some key features of DNA polymerases:

- They always need a template.

- They can only add nucleotides to the 3' end of a DNA strand.

- They can't start making a DNA chain from scratch, but require a pre-existing chain or short stretch of nucleotides called a primer.

- They proofread, or check their work, removing the vast majority of "wrong" nucleotides that are accidentally added to the chain.

The addition of nucleotides requires energy. This energy comes from the nucleotides themselves, which have three phosphates attached to them (much like the energy-carrying molecule ATP). When the bond between phosphates is broken, the energy released is used to form a bond between the incoming nucleotide and the growing chain.

In prokaryotes such as E. coli, there are two main DNA polymerases involved in DNA replication: DNA pol III (the major DNA-maker), and DNA pol I, which plays a crucial supporting role we'll examine later.

Starting DNA Replication

Replication always starts at specific locations on the DNA, which are called origins of replication and are recognized by their sequence.

E. coli, like most bacteria, has a single origin of replication on its chromosome. The origin is about 245 base pairs long and has mostly A/T base pairs (which are held together by fewer hydrogen bonds than G/C base pairs), making the DNA strands easier to separate.

Specialized proteins recognize the origin, bind to this site, and open up the DNA. As the DNA opens, two Y-shaped structures called replication forks are formed, together making up what's called a replication bubble. The replication forks will move in opposite directions as replication proceeds.

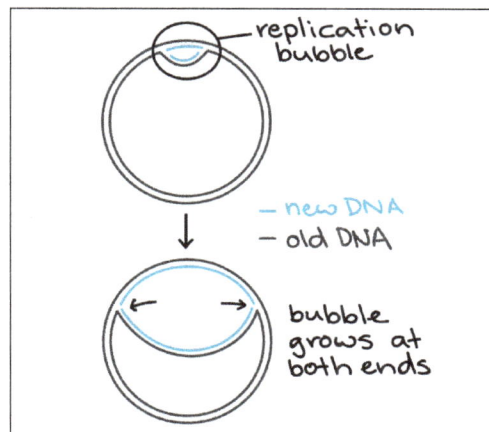

How does replication actually get going at the forks? Helicase is the first replication enzyme to load on at the origin of replication. Helicase's job is to move the replication forks forward by "unwinding" the DNA (breaking the hydrogen bonds between the nitrogenous base pairs).

Proteins called single-strand binding proteins coat the separated strands of DNA near the replication fork, keeping them from coming back together into a double helix.

Primers and Primase

DNA polymerases can only add nucleotides to the 3' end of an existing DNA strand. (They use the free -OH group found at the 3' end as a "hook," adding a nucleotide to this group in the polymerization reaction.) How, then, does DNA polymerase add the first nucleotide at a new replication fork.

The problem is solved with the help of an enzyme called primase. Primase makes an RNA primer, or short stretch of nucleic acid complementary to the template, that provides a 3' end for DNA polymerase to work on. A typical primer is about five to ten nucleotides long. The primer primes DNA synthesis, i.e., gets it started.

Once the RNA primer is in place, DNA polymerase "extends" it, adding nucleotides one by one to make a new DNA strand that's complementary to the template strand.

Leading and Lagging Strands

In E. coli, the DNA polymerase that handles most of the synthesis is DNA polymerase III. There are two molecules of DNA polymerase III at a replication fork, each of them hard at work on one of the two new DNA strands.

DNA polymerases can only make DNA in the 5' to 3' direction, and this poses a problem during replication. A DNA double helix is always anti-parallel; in other words, one strand runs in the 5' to 3' direction, while the other runs in the 3' to 5' direction. This makes it necessary for the two new strands, which are also antiparallel to their templates, to be made in slightly different ways.

One new strand, which runs 5' to 3' towards the replication fork, is the easy one. This strand is made continuously, because the DNA polymerase is moving in the same direction as the replication fork. This continuously synthesized strand is called the leading strand.

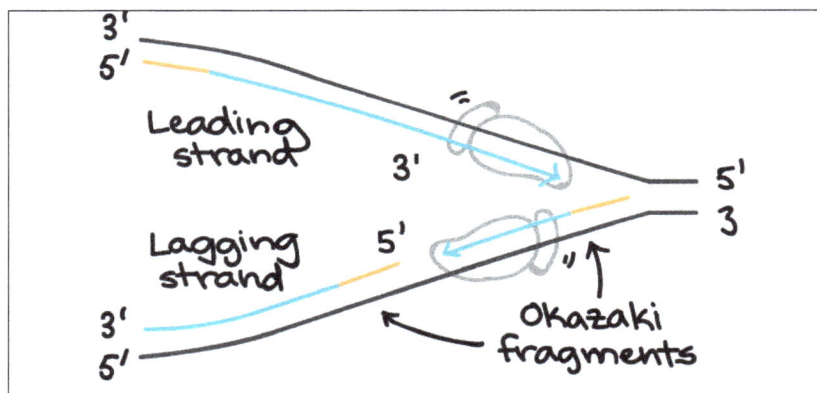

The other new strand, which runs 5' to 3' away from the fork, is trickier. This strand is made in fragments because, as the fork moves forward, the DNA polymerase (which is moving away from the fork) must come off and reattach on the newly exposed DNA. This tricky strand, which is made in fragments, is called the lagging strand.

The small fragments are called Okazaki fragments, named for the Japanese scientist who discovered them. The leading strand can be extended from one primer alone, whereas the lagging strand needs a new primer for each of the short Okazaki fragments.

Maintenance and Cleanup

Some other proteins and enzymes, in addition the main ones above, are needed to keep DNA replication running smoothly. One is a protein called the sliding clamp, which holds DNA polymerase

III molecules in place as they synthesize DNA. The sliding clamp is a ring-shaped protein and keeps the DNA polymerase of the lagging strand from floating off when it re-starts at a new Okazaki fragment. Topoisomerase also plays an important maintenance role during DNA replication. This enzyme prevents the DNA double helix ahead of the replication fork from getting too tightly wound as the DNA is opened up. It acts by making temporary nicks in the helix to release the tension, then sealing the nicks to avoid permanent damage.

Finally, there is a little cleanup work to do if we want DNA that doesn't contain any RNA or gaps. The RNA primers are removed and replaced by DNA through the activity of DNA polymerase I, the other polymerase involved in replication. The nicks that remain after the primers are replaced get sealed by the enzyme DNA ligase.

Polymerase Chain Reaction

The Polymerase Chain Reaction (PCR) is used to amplify specific regions of a DNA strand millions of times. A region may be a number of loci, a single gene, a part of a gene, or a non-coding sequence. This technique produces a useful quantity of DNA for analysis, be it medical, forensic or some other form of analysis. Amplification of DNA from as little as a single cell is possible. Whole genome amplification is also possible.

PCR utilizes a heat stable DNA polymerase, Taq polymerase (or Taq DNA polymerase), named after the thermophilic bacterium Thermus aquaticus, from which it was originally isolated. T. aquaticus is a bacterium that lives in hot springs and hydrothermal vents, and Taq polymerase is able to withstand the high temperatures required to denature DNA during PCR. Taq polymerase's optimum temperature for activity is between 75 °C and 80 °C. Recently other DNA polymerases have also been used for PCR.

A basic PCR involves a series of repeating cycles involving three main steps:

1. Denaturation of the double stranded DNA.

2. Annealing of specific oligonucleotide primers.

3. Extension of the primers to amplify the region of DNA of interest.

The oligonucleotide primers are single stranded pieces of DNA that correspond to the 5' and 3' ends of the DNA region to be amplified. These primers will anneal to the corresponding segment of denatured DNA. Taq Polymerase, in the presence of free deoxynucleotide triphosphates (dNTPs), will extend the primers to create double stranded DNA. After many cycles of denaturation, annealing and extension, the region between the two primers will be greatly amplified.

The PCR is commonly carried out in a thermal cycler, a machine that automatically allows heating and cooling of the reactions to control the temperature required at each reaction step. The PCR usually consists of a series of about 30 to 35 cycles. Most commonly, PCR is carried out in three repeating steps, with some modifications for the first and last step.

PCR is usually performed in small tubes or wells in a tray, each often beginning with the complete genome of the species being studied. As only a specific sequence from that genome is of interest, the sequence specific primers are targeted to that sequence. PCR is done with all the building blocks necessary to create DNA: template DNA, primers, dNTPs, and a DNA polymerase.

The three basic steps of PCR are:

- Denaturation step: This step is the first regular cycling event and consists of heating the reaction to 94 - 98 °C for 30 to 60 seconds. It disrupts the hydrogen bonds between complementary bases of the DNA strands, yielding single strands of DNA.

- Annealing step: The reaction temperature is lowered to 50-65 °C for 30 to 60 seconds, allowing annealing of the primers to the single-stranded DNA template. Stable hydrogen bonds form between the DNA strand (the template) and the primers when the primer sequence very closely matches the complementary template sequence. Primers are usually 17 - 22 nucleotides long and are carefully designed to bind to only one site in the genome. The polymerase binds to the primer-template hybrid and begins DNA synthesis.

- Extension step: A temperature of around 72 °C is used for this step, which is close to the optimum temperature of Taq polymerase. At this step the Taq polymerase extends the primer by adding dNTPs, using one DNA strand as a template to create a the other (new) DNA strand. The extension time depends on the length of the DNA fragment to be amplified. As a standard, at its optimum temperature, the DNA polymerase will polymerize a thousand bases in one minute.

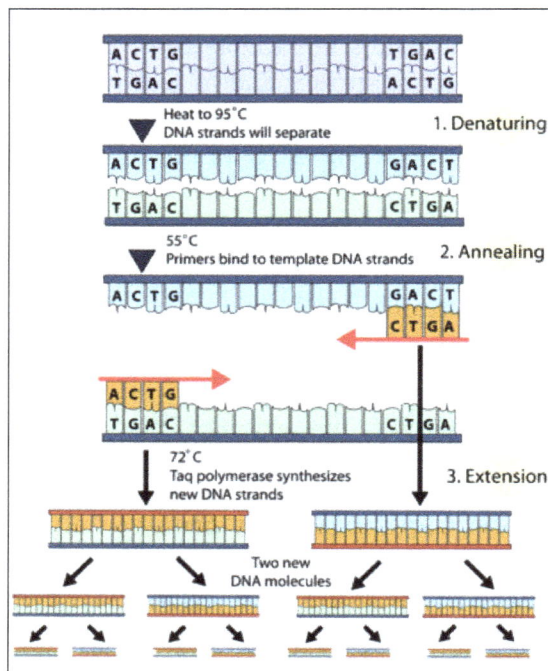

The Polymerase Chain Reaction. The polymerase chain reaction involves three steps. High temperatures are needed for the process to work. The enzyme Taq polymerase is used in step 3 because it can withstand high temperatures.

Utilizing PCR, DNA can be amplified millions of times to generate quantities of DNA that can be used for a number of purposes. These include the use of DNA for prenatal or genetic testing, such as testing for a specific mutation. PCR has revolutionized the fields of biotechnology, human genetics, and a number of other sciences.

Applications of Recombinant DNA Technology

The three important applications are: (1) Applications in Crop Improvement (2) Applications in Medicines and (3) Industrial Applications.

Applications in Crop Improvement

Genetic engineering has several potential applications in crop improvement, such as given below:

1. Distant Hybridization: With the advancement of genetic engineering, it is now possible to transfer genes between distantly related species. The barriers of gene transfer between species or even genera have been overcome. The desirable genes can be transferred even from lower organisms to higher organisms through recombinant DNA technology.

2. Development of Transgenic Plants: Genetically transformed plants which contain foreign genes are called transgenic plants. Resistance to diseases, insects and pests, herbicides, drought; metal toxicity tolerance; induction of male sterility for plant breeding purpose; and improvement of quality can be achieved through this recombinant DNA technology. BT-cotton, resistant to bollworms is a glaring example.

3. Development of Root Nodules in Cereal Crops: Leguminous plants have root-nodules which contain nitrogen fixing bacteria Rhizobium. This bacteria converts the free atmospheric nitrogen into nitrates in the root nodules. The bacterial genes responsible for this nitrogen fixation can be transferred now to cereal crops like wheat, rice, maize, barley etc. through the techniques of genetic engineering thus making these crops too capable of fixing atmospheric nitrogen.

4. Development of C_4 Plants: Improvement in yield can be achieved by improving the photosynthetic efficiency of crop plants. The photosynthetic rate can be increased by conversion of C_3 plants into C_4 plants, which can be achieved either through protoplasm fusion or recombinant DNA technology C_4 plants have higher potential rate of biomass production than C_3 plants. Most C_4 plants (sorghum, sugarcane, maize, some grasses) are grown in tropical and subtropical zones.

Applications in Medicines

Biotechnology, especially genetic engineering plays an important role in the production of antibiotics, hormones, vaccines and interferon in the field of medicines.

1. Production of Antibiotics: Penicillium and Streptomyces fungi are used for mass production of famous antibiotics penicillin and streptomycin. Genetically efficient strains of these fungi have been developed to greatly increase the yield of these antibiotics.

2. Production of Hormone Insulin: Insulin, a hormone, used by diabetics, is usually extracted from pancreas of cows and pigs. This insulin is slightly different in structure

from human insulin. As a result, it leads to allergic reactions in about 5% patients. Human gene for insulin production has been incorporated into bacterial DNA and such genetically engineered bacteria are used for large scale production of insulin. This insulin does not cause allergy.

3. Production of Vaccines: Vaccines are now produced by transfer of antigen coding genes to disease causing bacteria. Such antibodies provide protection against the infection by the same bacteria or virus.

4. Production of Interferon: Interferon's are virus-induced proteins produced by virus-infected cells. Interferon are antiviral in action and act as first line of defense against viruses causing serious infections, including breast cancer and lymph nodes malignancy. Natural interferon is produced in very small quality from human blood cells. It is thus very costly also. It is now possible to produce interferon by recombinant DNA technology at much cheaper rate.

5. Production of Enzymes: Some useful enzymes can also be produced by recombinant DNA technique. For instance, enzyme urikinase, which is used to dissolve blood clots, has been produced by genetically engineered microorganisms.

6. Gene Therapy: Genetic engineering may one day enable the medical scientists to replace the defective genes responsible for hereditary diseases (e.g., haemophilia, phenylketonuria, alkaptonuria) with normal genes. This new system of therapy is called gene therapy.

7. Solution of Disputed Parentage: Disputed cases of parentage can now be solved most accurately by recombinant technology than by blood tests.

The following table shows some medically useful recombinant products and their applications:

Medically useful Recombinant products	Application
Human inslin	Treatment of insulin-dependent diabetes
Human growth hormone	Replacement of missing hormone in short stature people
Calcitonin	Treatment of rickets
	Treatment of infertility
Chronic gonadotropin	Replacement of clotting factor missing in patients with haemophilia A/B
Blood clotting factor viii/ix	Dissolving of blood clots after heart attacks and strokes
Platelet derived growth factor	Stimulation of the formation of erythrocytes (RBCs) for patients suffering from anaemia during kidney dialysis or side effects of stimulation of wound healing.
Interferon	Treatment of pathogenic viral infections, cancer.
Interleukins	Enhancement of action of immune system
Vaccines	Prevention of infectious diseases such as hepatitis B, herpes, influenza, pertussis, meningitis, etc.

8. Diagnosis of Disease: Recombinant DNA technology has provided a broad range of tools to help physicians in the diagnosis of diseases. Most of these involve the construction of

probes: short Segments of single stranded DNA attached to a radioactive or fluorescent marker. Such probes are now used for identification of infectious agents, for instance, food poisoning Salmonella, Pus forming Staphylococcus, hepatitis virus, HIV, etc. By testing the DNA of prospective genetic disorder carrier parents, their genotype can be determined and their chances of producing an afflicted child can be predicted.

9. Production of Transgenic Animals: Animals which carry foreign genes are called transgenic animals.

Examples: Cow, sheep, goat – therapeutic; human proteins in their milk. Fish like common carp, cat fish, salmon and gold fish contain human growth hormone (hGH).

Industrial Applications

In industries, recombinant DNA technique will help in the production of chemical compounds of commercial importance, improvement of existing fermentation processes and production of proteins from wastes. This can be achieved by developing more efficient strains of microorganisms. Specially developed microorganisms may be used even to clean up the pollutants. Thus, biotechnology, especially recombinant DNA technology has many useful applications in crop improvement, medicines and industry.

References

- Recombinant-DNA-technology, science: britannica.com, Retrieved 7 May, 2019

- Recombinant-dna-technology-with-diagram, recombinant-dna-technology, dna: biologydiscussion.com, Retrieved 18 April, 2019

- Overview-dna-cloning, dna-cloning-tutorial, biology, science: khanacademy.org, Retrieved 14 June, 2019

- Gene-cloning-and-vectors-definition-and-major-types: thebalance.com, Retrieved 28 February, 2019

- 3-main-classification-of-vectors-with-diagram, recombinant-dna-technology: biotechnologynotes.com, Retrieved 19 May, 2019

- Dna-sequencing, dna-sequencing-pcr-electrophoresis, biotech-dna-technology, biology, science: khanacademy.org, Retrieved 28 April, 2019

- Molecular-mechanism-of-dna-replication, dna-replication, dna-as-the-genetic-material, biology, science: khanacademy.org, Retrieved 1 August, 2019

- Applications-of-recombinant-dna-technology-3-applications, recombinant-dna-technology, dna: biologydiscussion.com, Retrieved 13 July, 2019

Genetic Engineering

The direct manipulation of an organism's DNA to change its characteristics in a specific manner is known as genetic engineering. Medicine, industrial biotechnology and agriculture are the various fields where genetic engineering is applied. The diverse applications of genetic engineering in the current scenario have been thoroughly discussed in this chapter.

Genetic engineering is the direct altering of an organism's genome. This is achieved through manipulation of the DNA. Doing this is possible because DNA is like a universal language; all DNA for all organisms is made up of the same nucleotide building blocks. Thus, it is possible for genes from one organism to be read by another organism. In the cookbook analogy, this equates to taking a recipe from one organism's cookbook and putting into another cookbook. Now imagine that all cookbooks are written in the same language; thus, any recipe can be inserted and used in any other cookbook.

In practice, since DNA contains the genes to build certain proteins, by changing the DNA sequence, engineers are able to provide a new gene for a cell/organism to create a different protein. The new instructions may supplement the old instructions such that an extra trait is exhibited, or they may completely replace the old instructions such that a trait is changed.

Genetic Engineering Technique

The process for genetic engineering begins the same for any organism being modified:

1. Identify an organism that contains a desirable gene.

2. Extract the entire DNA from the organism.

3. Remove this gene from the rest of the DNA. One way to do this is by using a restriction enzyme. These enzymes search for specific nucleotide sequences where they will "cut" the DNA by breaking the bonds at this location.

4. Insert the new gene to an existing organism's DNA. This may be achieved through a number of different processes.

When modifying bacteria, the most common method for this final step is to add the isolated gene to a plasmid, a circular piece of DNA used by bacteria. This is done by "cutting" the plasmid with the same restriction enzyme that was used to remove the gene from the original DNA. The new gene can now be inserted into this opening in the plasmid and the DNA can be bonded back together using another enzyme called ligase. This process, creates a recombinant plasmid. In this case, the recombinant plasmid is also referred to as a bacterial artificial chromosome (BAC).

Once the recombinant DNA has been built, it can be passed to the organism to be modified. If modifying bacteria, this process is quite simple. The plasmid can be easily inserted into the bacteria where the bacteria treat it as their own DNA. For plant modification, certain bacteria such as Agrobacterium tumefaciens may be used because these bacteria permit their plasmids to be passed to the plant's DNA.

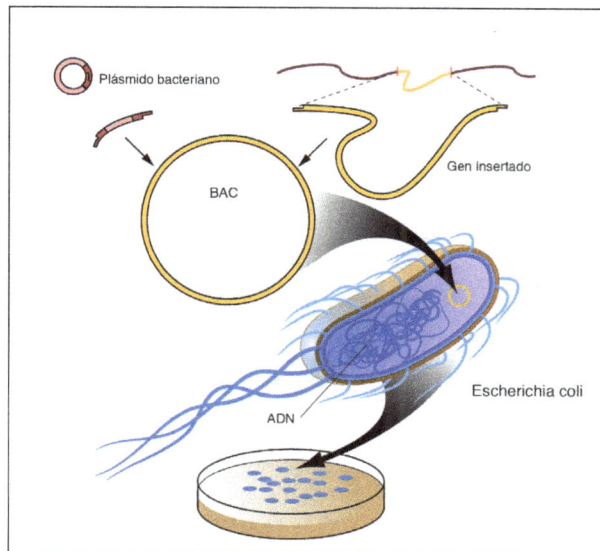

Building a recombinant plasmid to modify bacteria.

Genetically Modified Organism

Genetically modified organism (GMO) is an organism whose genome has been engineered in the laboratory in order to favour the expression of desired physiological traits or the production of desired biological products. In conventional livestock production, crop farming, and even pet breeding, it has long been the practice to breed select individuals of a species in order to produce offspring that have desirable traits. In genetic modification, however, recombinant genetic technologies are employed to produce organisms whose genomes have been precisely altered at the molecular level, usually by the inclusion of genes from unrelated species of organisms that code for traits that would not be obtained easily through conventional selective breeding.

GMOs are produced through using scientific methods that include recombinant DNA technology and reproductive cloning. In reproductive cloning, a nucleus is extracted from a cell of the individual to be cloned and is inserted into the enucleated cytoplasm of a host egg. The process results in the generation of an offspring that is genetically identical to the donor individual. The first animal produced by means of this cloning technique with a nucleus from an adult donor cell (as opposed to a donor embryo) was a sheep named Dolly, born in 1996. Since then a number of other animals, including pigs, horses, and dogs, have been generated by reproductive cloning technology. Recombinant DNA technology, on the other hand, involves the insertion of one or more individual genes from an organism of one species into the DNA (deoxyribonucleic acid) of another. Whole-genome replacement, involving the transplantation of one bacterial genome into the "cell body," or cytoplasm, of another microorganism, has been reported, although this technology is still limited to basic scientific applications.

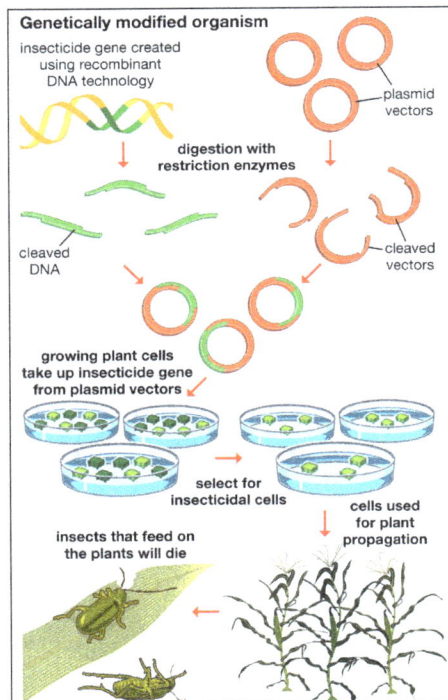

Genetically modified organisms are produced using scientific methods that
include recombinant DNA technology.

GMOs produced through genetic technologies have become a part of everyday life, entering into
society through agriculture, medicine, research, and environmental management. However, while
GMOs have benefited human society in many ways, some disadvantages exist; therefore, the production of GMOs remains a highly controversial topic in many parts of the world.

Genetically modified (GM) foods were first approved for human consumption in the United States
in 1994, and by 2014–15 about 90 percent of the corn, cotton, and soybeans planted in the United
States were GM. By the end of 2010, GM crops covered more than 10 million square kilometres
(3.86 million square miles) of land in 29 countries worldwide—one-tenth of the world's farmland.
The majority of GM crops were grown in the Americas.

Genetically modified corn (maize).

Engineered crops can dramatically increase per area crop yields and, in some cases, reduce the use of chemical insecticides. For example, the application of wide-spectrum insecticides declined in many areas growing plants, such as potatoes, cotton, and corn, that were endowed with a gene from the bacterium Bacillus thuringiensis, which produces a natural insecticide called Bt toxin. Field studies conducted in India in which Bt cotton was compared with non-Bt cotton demonstrated a 30–80 percent increase in yield from the GM crop. This increase was attributed to marked improvement in the GM plants' ability to overcome bollworm infestation, which was otherwise common. Studies of Bt cotton production in Arizona, U.S., demonstrated only small gains in yield—about 5 percent—with an estimated cost reduction of $25–$65 (USD) per acre owing to decreased pesticide applications. In China, where farmers first gained access to Bt cotton in 1997, the GM crop was initially successful. Farmers who had planted Bt cotton reduced their pesticide use by 50–80 percent and increased their earnings by as much as 36 percent. By 2004, however, farmers who had been growing Bt cotton for several years found that the benefits of the crop eroded as populations of secondary insect pests, such as mirids, increased. Farmers once again were forced to spray broad-spectrum pesticides throughout the growing season, such that the average revenue for Bt growers was 8 percent lower than that of farmers who grew conventional cotton. Meanwhile, Bt resistance had also evolved in field populations of major cotton pests, including both the cotton bollworm (Helicoverpa armigera) and the pink bollworm (Pectinophora gossypiella).

Other GM plants were engineered for resistance to a specific chemical herbicide, rather than resistance to a natural predator or pest. Herbicide-resistant crops (HRC) have been available since the mid-1980s; these crops enable effective chemical control of weeds, since only the HRC plants can survive in fields treated with the corresponding herbicide. Many HRCs are resistant to glyphosate (Roundup), enabling liberal application of the chemical, which is highly effective against weeds. Such crops have been especially valuable for no-till farming, which helps prevent soil erosion. However, because HRCs encourage increased application of chemicals to the soil, rather than decreased application, they remain controversial with regard to their environmental impact. In addition, in order to reduce the risk of selecting for herbicide-resistant weeds, farmers must use multiple diverse weed-management strategies.

Another example of a GM crop is "golden" rice, which originally was intended for Asia and was genetically modified to produce almost 20 times the beta-carotene of previous varieties. Golden rice was created by modifying the rice genome to include a gene from the daffodil Narcissus pseudonarcissus that produces an enzyme known as phyotene synthase and a gene from the bacterium Erwinia uredovora that produces an enzyme called phyotene desaturase. The introduction of these genes enabled beta-carotene, which is converted to vitamin A in the human liver, to accumulate in the rice endosperm—the edible part of the rice plant—thereby increasing the amount of beta-carotene available for vitamin A synthesis in the body. In 2004 the same researchers who developed the original golden rice plant improved upon the model, generating golden rice 2, which showed a 23-fold increase in carotenoid production.

Another form of modified rice was generated to help combat iron deficiency, which impacts close to 30 percent of the world population. This GM crop was engineered by introducing into the rice genome a ferritin gene from the common bean, Phaseolus vulgaris, that produces a protein capable of binding iron, as well as a gene from the fungus Aspergillus fumigatus that produces an enzyme capable of digesting compounds that increase iron bioavailability via digestion of phytate (an inhibitor of iron absorption). The iron-fortified GM rice was engineered to overexpress an existing rice gene that produces a cysteine-rich metallothioneinlike (metal-binding) protein that enhances iron absorption.

A variety of other crops modified to endure the weather extremes common in other parts of the globe are also in production.

GMOs in Medicine and Research

GMOs have emerged as one of the mainstays of biomedical research since the 1980s. For example, GM animal models of human genetic diseases enabled researchers to test novel therapies and to explore the roles of candidate risk factors and modifiers of disease outcome. GM microbes, plants, and animals also revolutionized the production of complex pharmaceuticals by enabling the generation of safer and cheaper vaccines and therapeutics. Pharmaceutical products range from recombinant hepatitis B vaccine produced by GM baker's yeast to injectable insulin (for diabetics) produced in GM Escherichia coli bacteria and to factor VIII (for hemophiliacs) and tissue plasminogen activator (tPA, for heart attack or stroke patients), both of which are produced in GM mammalian cells grown in laboratory culture. Furthermore, GM plants that produce "edible vaccines" are under development. An edible vaccine is an antigenic protein that is produced in the consumable parts of a plant (e.g., fruit) and absorbed into the bloodstream when the parts are eaten. Once absorbed into the body, the protein stimulates the immune system to produce antibodies against the pathogen from which the antigen was derived. Such vaccines could offer a safe, inexpensive, and painless way to provide vaccines, particularly in less-developed regions of the world, where the limited availability of refrigeration and sterile needles has been problematic for some traditional vaccines. Novel DNA vaccines may be useful in the struggle to prevent diseases that have proved resistant to traditional vaccination approaches, including HIV/AIDS, tuberculosis, and cancer.

Genetic modification of insects has become an important area of research, especially in the struggle to prevent parasitic diseases. For example, GM mosquitoes have been developed that express a small protein called SM1, which blocks entry of the malaria parasite, Plasmodium, into the mosquito's gut. This results in the disruption of the parasite's life cycle and renders the mosquito malaria-resistant. Introduction of these GM mosquitoes into the wild could help reduce transmission of the malaria parasite. In another example, male Aedes aegypti mosquitoes engineered with a method known as the sterile insect technique transmit a gene to their offspring that causes the offspring to die before becoming sexually mature. In field trials in a Brazil suburb, A. aegypti populations declined by 95 percent following the sustained release of sterile GM males.

Finally, genetic modification of humans via gene therapy is becoming a treatment option for diseases ranging from rare metabolic disorders to cancer. Coupling stem cell technology with recombinant DNA methods allows stem cells derived from a patient to be modified in the laboratory to introduce a desired gene. For example, a normal beta-globin gene may be introduced into the DNA of bone marrow-derived hematopoietic stem cells from a patient with sickle cell anemia; introduction of these GM cells into the patient could cure the disease without the need for a matched donor.

Role of GMOs in Environmental Management

Another application of GMOs is in the management of environmental issues. For example, some bacteria can produce biodegradable plastics, and the transfer of that ability to microbes that can be easily grown in the laboratory may enable the wide-scale "greening" of the plastics industry. In the early 1990s, Zeneca, a British company, developed a microbially produced biodegradable plastic called Biopol (polyhydroxyalkanoate, or PHA). The plastic was made with the use of a GM bacterium, Ralstonia eutropha, to convert glucose and a variety of organic acids into a flexible polymer. GMOs endowed with the bacterially encoded ability to metabolize oil and heavy metals may provide efficient bioremediation strategies.

Gene Libraries

A DNA library is a set of cloned fragments that collectively represent the genes of a particular organism. Particular genes can be isolated from DNA libraries, much as books can be obtained from conventional libraries.

The secret is knowing where and how to look. There are two general types of gene library: a genomic library, which consists of the total chromosomal DNA of an organism; and a cDNA library, which represents the mRNA from a cell or tissue at a specific point of time.

The choice of the particular type of gene library depends on a number of factors, the most important being the final application of any DNA fragment derived from the library. If the ultimate aim understands the control of protein production for a particular gene or its architecture, then genomic libraries must be used.

However, if the goal is the production of new or modified proteins, or the determination of tissue-specific expression of timing patterns, cDNA libraries are more appropriate. The main consideration in the construction of genomic or cDNA libraries is, therefore, the nucleic acid starting material. Since the genome of an organism is fixed, chromosomal DNA may be isolated from almost any cell type in order to prepare genomic DNA.

In contrast, however, cDNA libraries represent only mRNA being produced from a specific cell type at a particular time in the cell's development. Thus, it is important to consider carefully the cell or tissue type from which the mRNA is to be deriver in the construction of cDNA libraries.

There are a variety of cloning vectors available, many based on naturally occurring molecules such as bacterial plasmids or bacteria-infecting viruses. The choice of vector also depends on whether a genomic library or cDNA library is constructed.

Constructing Gene Libraries

Digesting Genomic DNA Molecules

After genomic DNA has been isolated and purified, it is digested with restriction endonucleases. These enzymes are the key to molecular cloning because of the specificity they have for particular DNA sequences. It is important to note that every copy of a given DNA molecule from a specific organism will give the same set of fragments when digested with a particular enzyme.

DNA from different organisms will, in general, give different sets of fragments when treated with the same enzyme. By digesting complex genomic DNA from an organism it is possible to repro-ducibly divide its genome into a large number of small fragments, each approximately the size of a single gene. Some enzymes cut straight across the DNA to give flush or blunt ends.

Other restriction enzymes make staggered single-strand cuts, producing short single-stranded projections at each end of the digested DNA. These ends are not only identical but complementary and will base-pair with each other; they are, therefore, known as cohesive or sticky ends. In addi-tion, the 5'-end projection of the DNA always retains the phosphate groups.

Over 500 restriction enzymes, recognizing more than 200 different sites, have been characterized. The choice of which enzyme to use depends on a number of factors. For example, the recognition sequence of 6 bp will occur, on average, every 4096 (46) bases, assuming a random sequence of each of the four bases.

This means that digesting genomic DNA with EcoRI, which recognizes the sequence 5'-GAAT-TC-3', will produce fragments each of which is, on average, just over 4 kb. Enzymes with 8 bp recognition sequences produce much longer fragments. Therefore, very large genomes, such as human DNA, are usually digested with enzymes that produce long DNA fragments. This makes subsequent steps more manageable, since a smaller number of those fragments need to be cloned and subsequently analyzed.

Ligating DNA Molecules

The DNA products resulting from restriction digestion to form sticky ends may be joined to any other DNA fragments treated with the same restriction enzyme. Thus, when the two sets of frag-ments are mixed; base-pairing between sticky ends will result in the annealing of fragments that were derived from different starting DNA. There will, of course, also be pairing of fragments de-rived from the same starting DNA molecules, termed re-annealing.

All these pairing are transient, owing to the weakness of hydrogen bonding between the few bases in the sticky ends, but they can be stabilized by use of an enzyme, DNA ligase, in a process termed liga-tion. This enzyme, usually isolated from bacteriophage T4 and called T4 DNA ligase, forms a covalent bond between the 5'- phosphate at the end of one strand and the 3'-hydroxyl of the adjacent strand.

The reaction which is ATP dependent is often carried out at 10 °C to lower the kinetic energy of the molecules, and so reduce the chances of base-paired sticky ends parting before they have been stabilized by ligation. However, long reaction times are needed to compensate for the low activity of DNA ligase in the cold. It is also possible to join blunt ends of DNA molecules, although the ef-ficiency of this reaction is much lower than in sticky-ended ligations.

Since ligation reconstructs the site of cleavage, recombinant molecules produced by ligation of sticky ends can be cleaved again at the 'joins', using the same restriction enzyme that was used to generate the fragments initially. In order to propagate digested DNA from an organism it is neces-sary to join or ligate that DNA with a specialized DNA carrier molecule termed a vector.

Each DNA fragment is inserted by ligation into vector DNA molecule, which allows the whole recombinant DNA to then be replicated indefinitely within microbial cells. In this way a DNA fragment can be cloned to provide sufficient material for further detailed analysis or for further manipulations. Thus, all of the DNA extracted from an organism and digested with a restriction enzyme will result in a collection of clones. This collection of clones is known as a gene library.

Genomic Libraries

Any particular gene constitutes only a small part of an organism's genome. For example, if the organism is a mammal whose entire genome encompasses some 106 kbp and the gene is 10 kbp, then the gene represents only 0.001% of the total nuclear DNA. It is impractical to attempt to recover such rare sequences directly from isolated nuclear DNA because of the overwhelming amount of extraneous DNA sequences.

Instead, a genomic library is prepared by isolating total DNA from the organism, digesting it into fragments of suitable size, and cloning the fragments into an appropriate vector. This approach is called shotgun cloning because the strategy has no way of targeting a particular gene but instead seeks to clone all the genes of the organism at one time.

The intent is that at least one recombinant clone will contain at least part of the gene of interest. This can be achieved by partial restriction digestion with an enzyme that recognizes tetra nucleotide sequences. Complete digestion with such an enzyme would produce a large number of very short fragments, but, if the enzyme is allowed to cleave only a few of its potential restriction sites before the reaction is stopped, each DNA molecule will be cut into relatively large fragments.

Average fragment size will depend on the relative concentrations of DNA and restriction enzyme and, in particular, on the conditions and durations of incubation. It is also possible to produce fragments of DNA by physical shearing, although the ends of the fragments may need to be repaired to make them flush ended. This is achieved by using a modified DNA polymerase termed Klenow polymerase.

This is prepared by cleavage of DNA polymerase with subtilizing, giving a large enzyme fragment which has no $5' \rightarrow 3'$ exonuclease activity, but which still acts as $5' \rightarrow 3'$ polymerase. Using the appropriate dNTPs, this will fill in any recessed $3'$ ends on the sheared DNA. The mixture of DNA fragments is then ligated with a vector, and subsequently cloned.

If enough clones are produced there will be a very high chance that any particular DNA fragment, such as a gene, will be present in at least one of the clones. To keep the number of clones to a manageable size, fragments about 10 kb in length are needed for prokaryotic libraries, but the length must be increased to about 40 kb for mammalism libraries.

Genomic libraries have been prepared from hundreds of different species. Many clones must be created to be confident that the genomic library contains the gene of interest. The probability, P, that some number of clones, N, contains a particular fragment representing a fraction, f, of the genome is,

$$P = 1 - (1 - f)^N.$$
$$\text{Thus, } N = \ln(1 - P) / \ln(1 - f).$$

For example, if the library consists of 10 kbp fragments of the E. coli genome (4640 kbp total), over

2000 individual clones must be screened to have a 99% probability (P = 0.99) of finding a particular fragment. Since/ =10/4640 = 0.0022 and P – 0.99, N = 2093. For a 99% probability of finding a particular sequence within the 3 x 106 kbp human genome, N would equal almost 1.4 million if the cloned fragments averaged 10 kbp in size. The need for cloning vectors capable of carrying very large DNA inserts becomes obvious from these numbers.

Combinatorial Libraries

Specific recognition and binding of other molecules is a defining characteristic of any protein or nucleic acid. Often, target ligands of a particular protein are unknown, or, in other instances, a unique ligand for a known protein may be sought in the hope of blocking the activity of the protein or otherwise perturbing its function.

Combinatorial libraries are the products of emerging strategies to facilitate the identification and characterization of possible ligands for a protein. These strategies are also applicable to the study of nucleic acids. Unlike genomic libraries, combinatorial libraries consist of synthetic oligomers. Arrays of synthetic oligonucleotides printed as tiny dots on miniature solid supports are known as DNA chips.

Specifically, combinatorial libraries contain very large numbers of chemically synthesized molecules (such as peptides or oligonucleotides) with randomized sequences or structures. Such libraries are designed and constructed with the hope that one molecule among a vast number will be recognized as a ligand by the protein (or nucleic acid) of interest.

If so, perhaps that molecule will be useful in a pharmaceutical application, for instance as a drug to treat a disease involving the protein to which it binds. An example of this strategy is the preparation of a synthetic combinatorial library of hexapeptides. The maximum number of sequence combinations for hexapeptides is 206 or 64,000,000.

One approach to simplify preparation and screening possibilities for such a library is to specify the first two amino acids in the hexapeptide while the next four are randomly chosen. In this approach, 400 libraries (202) are synthesized, each of which is unique in terms of the amino acids at positions 1 and 2 but random at the other four positions (as in AAXXXX, ACXXXX, ADXXXX, etc.) so that each of the 400 libraries contains 204 or 1,60,000 different sequence combinations.

Screening these libraries with the protein of interest reveals which of the 400 libraries contains a ligand with high affinity. This library is then systematically expanded by specifying the first 3 amino acids (knowing from the chosen 1-of-400 libraries which amino acids are best as the first 2); only 20 synthetic libraries (each containing 203 or 8000 hexapeptides) are made here (one for each third-position possibility, the remaining three positions being randomized).

Selection for ligand binding, again with the protein of interest, reveals the best of these 20, and this particular library is then varied systematically at the fourth position, creating 20 more libraries (each containing 202 or 400 hexapeptides). This cycle of synthesis, screening, and selection is repeated until all six positions in the hexapeptide are optimized to create the best ligand for the protein.

A variation on this basic strategy using synthetic oligonucleotides rather than peptides identified a unique 15-mer (sequence GGTTGGTGTGGTTGG) with high affinity (KD = 2.7 nM) toward thrombin, a serine protease in the blood coagulation pathway. Thrombin is a major target for the pharmacological prevention of clot formation in coronary thrombosis.

Screening Libraries

A common method of screening plasmid-based genomic libraries is to carry out a colony hybridization experiment. The protocol is similar for phage-based libraries except that bacteriophage plaques, not bacterial colonies, are screened. In a typical experiment, host bacteria containing either a plasmid based or bacteriophage-based library are plated out on a petri dish and allowed to grow overnight to form colonies (or in the case of phage libraries, plaques).

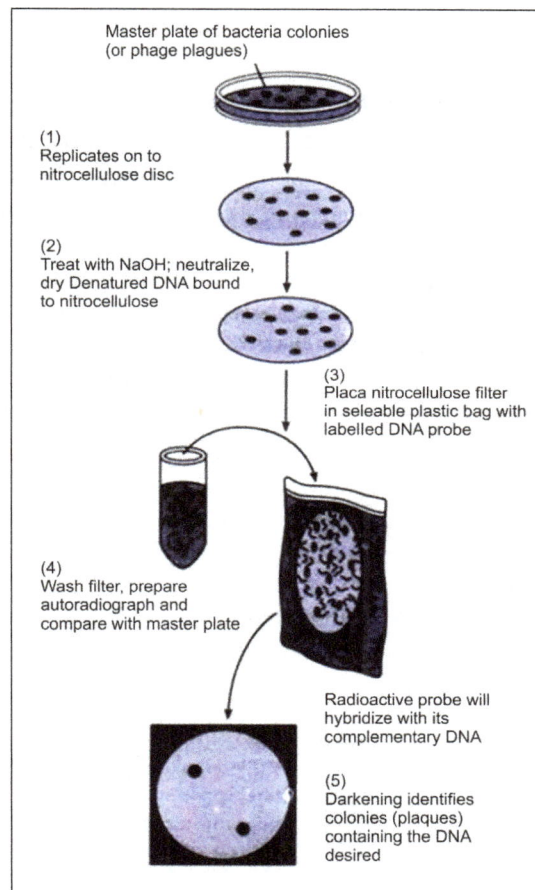

Screening a genomic library by colony hybridization (or plaque hybridization . Host bacteria tre. nslormed with a plasirid-based genOrniC library or infected with a bacteriophage-based gencierk library are Mated r1 a path plate and incubated overnight to allow bacterial doronies (or phage plaques) to form. A replica of the bacterial colonies (or plaques) is then obtained by overlaying the plate with a nitrocellulose disc (1) tarocellulos.a strongly binds nucleic acids: single-stranded nucleic acids are bound more tightly than double-stranded nuelOIC OCAS. (Nylon r-riembranes with similar nucleic acid y nd protein-bireing properties ere alp used.) Once the nitrocellurose disc has taken up an impression of the bacterial colonies NT plaques). it is removed and the petri plate is set aside and saved. The disc is treated with 2 M neutralized, and dried. (2) NaOH both lyse any bacleria (or phage particles) and dissociates the DNA strands. When the disc is dried, the DNA strands. become immobilized On the fitter. The dried disc is placed in a sealable plastic bag. and a sotution containing heat denatUrod (singte-alianded). labelled probe 3s added. (3) The bag is incubefeCr 10 allow annealing of the probe DNA to any target DNA sequences that might be present on the nitrocellulose. The biter is then washed, driedr and pieced on a piece of X-ray fiirn to obtain an mit:radiogram. (4) The position of any spots on the X-ray film reveals where the labelled probe has hybridized with target DNA. (5) The location of these spots can be Ltsed to recover the gendmic clone frown the bacteria (or Wagues) on tha original pain plate.

A replica of the bacterial colonies (or plaques) is then obtained by overlaying the plate with a nitrocellulose disc. The disc is removed, treated with alkali to dissociate bound DNA duplexes into single-stranded DNA, dried, and placed in a sealed bag with labelled probe. If the probe DNA is duplex DNA, it must be denatured by heating at 70 °C.

The probe and target DNA complementary sequences must be in a single stranded form if they are to hybridize with one another. Any DNA sequences complementary to probe DNA will be revealed by autoradiography of the nitrocellulose disc. Bacterial colonies (phage plaques) containing clones bearing target DNA are identified on the film and can be recovered from the master plate.

Probes for Southern Hybridization

Clearly, specific probes are essential reagents if the goal is to identify a particular gene against a background of innumerable DNA sequences. Usually, the probes that are used to screen libraries are nucleotide sequences that are complementary to some part of the target gene. To make useful probes requires some information about the gene's nucleotide sequence.

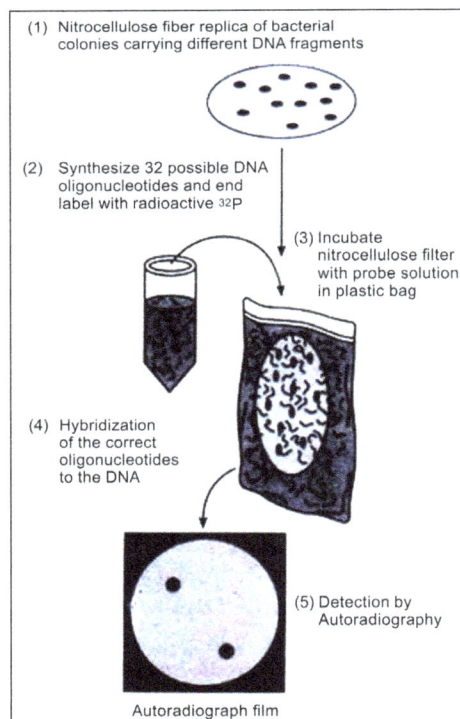

(1) Nitrocellulose fiber replica of bacterial colonies carrying different DNA fragments

(2) Synthesize 32 possible DNA oligonucleotides and end label with radioactive 32P

(3) Incubate nitrocellulose filter with probe solution in plastic bag

(4) Hybridization of the correct oligonucleotides to the DNA

(5) Detection by Autoradiography

Autoradiograph film

Sequence: A radioactively labelled set of DNA (degenerate) oligonucleotides representing all possible mRNA coding sequences is synthesized. (in this case, there are 25 or 32.)The complete mixture is used to probe the genomic library by colony hybridization.

Sometimes such information is available. Alternatively, if the amino acid sequence of the protein encoded by the gene is known, it is possible to work backward through the genetic code to the DNA sequence. Because the genetic code is degenerate (that is, several codons may specify the same amino acid), probes designed by this approach are usually degenerate oligonucleotides about 17 to 50 residues long (such oligonucleotides are so-called 17- to 50- mers).

The oligonucleotides are synthesized so that different bases are incorporated at sites where degeneracies occur in the codons. The final preparation thus consists of a mixture of equal-length oligonucleotides whose sequences vary to accommodate the degeneracies. Presumably, one oligonucleotide sequence in the mixture will hybridize with the target gene. These oligonucleotide probes are at least 17-mers because shorter degenerate oligonucleotides might hybridize with sequences unrelated to the target sequence.

A piece of DNA from the corresponding gene in a related organism can also be used as a probe in screening a library for a particular gene. Such probes are termed heterologous probes because they are not derived from the homologous (same) organism. Problems arise if a complete eukaryotic gene is the cloning target; eukaryotic genes can be tens or even hundreds of kilo-base pairs in size.

Genes of this size are fragmented in most cloning procedures. Thus, the DNA identified by the probe may represent a clone that carries only part of the desired gene. However, most cloning strategies are based on a partial digestion of the genomic DNA, a technique that generates an overlapping set of genomic fragments.

This being so, DNA segments from the ends of the identified clone can now be used to probe the library for clones carrying DNA sequences that flanked the original isolate in the genome. Repeating this process ultimately yields the complete gene among a subset of overlapping clones.

cDNA Libraries

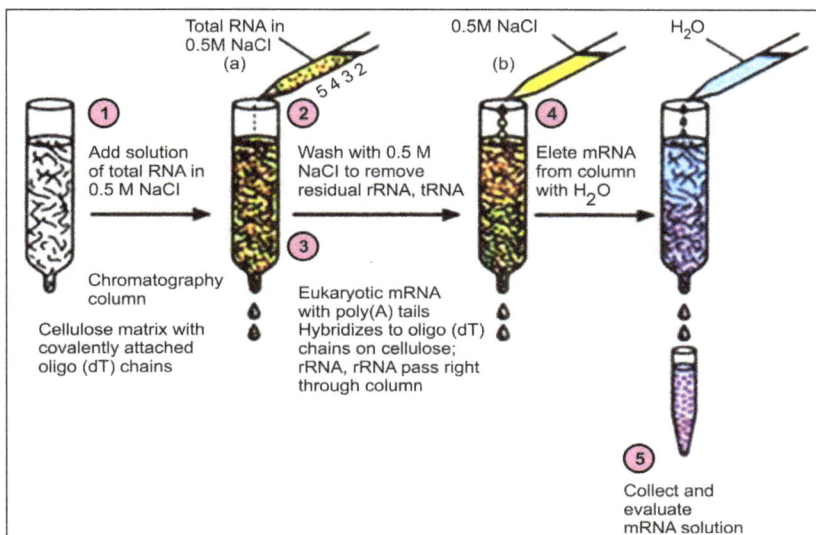

cDNAs are DNA molecules copied from mRNA templates. cDNA libraries are constructed by synthesizing cDNA from purified cellular mRNA. These libraries present an alternative strategy for gene isolation, especially eukaryotic genes. Because most eukaryotic mRNAs carry 3'-poly(A) tails, mRNA can be selectively isolated from preparations of total cellular RNA by oligo(dT)-cellulose chromatography. DNA copies of the purified mRNAs are synthesized by first annealing short oligo (dT) chains to the poly(A) tails.

Isolation of eukarybtic mRNA via oligo (dT)-cellulose chromatography. (a) In the preseinca of 0.5 NIaCI, the poly (A) tails f eukaryotic mRNA anneal with short oligo (dT) chains covalently attached to an insoluble chromatographic rnalrix such as cellulose. Other RNAs, such as rRNA (green), pass right through the chromalography column. (b) The column is washed with more 0.5 M NaCl to remove residual contaminants. (c) Then the poly(A) mRNA is recovered by washing the column with water because the base pairs formed between the poly (A) tails of the mRNA and the oligo (dT) chains are unstable in solution of low ionic strength.

These oligo(dT) chains serve as primers for reverse transcriptase-driven synthesis of DNA. (Random oligonucleotides can also be used as primers, with the advantages being less dependency on poly(A) tracts and increased likelihood of creating clones representing the 5′-ends of mRNAs.) Reverse transcriptase is an enzyme that synthesizes a DNA strand, copying RNA as the template. DNA polymerase is then used to copy the DNA strand and form a double-stranded (duplex DNA) molecule.

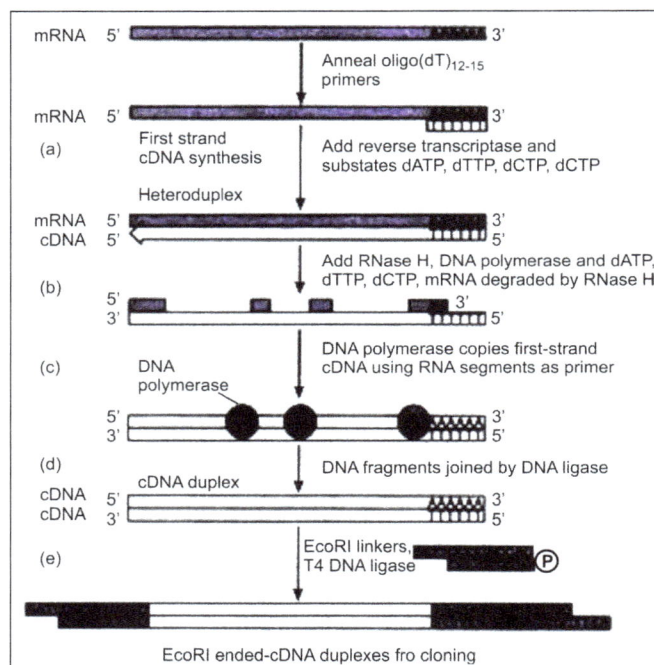

Reverse trenscriptese-driven synthesis of cDNA from oligo(dT) primers annealed to the poly(A) tails of purified eukaryotic mRNA. (a) Oligo(dT) chains serve as primers for synthesis of a DNA copy of the mRNA by reverse transcriptase. Following completion of first-strand cDNA synthesis by reverse trenscriptase, RNase H and DNA polymerase are added (b). RNase H specifically digests RNA strands in DNA:RNA hybrid duplexes. DNA polymerase copies the first-strand cDNA, using as primers the residual RNA segments after RNase H has created nicks and gaps (c). DNA polymerase has a 5′—3′ exonuclease activity that removes the residual RNA as it fills in with DNA. The nicks remaining in the second-strand DNA are sealed by DNA ligase (d), yielding duplex cDNA. EcoRI adapters with 5′-overhangs are then ligated onto the cDNA duplexes (e) using phage T4 DNA ligase to create EcoRI ended cDNA for insertion into cloning vector.

Ligation of blunt-ended DNA fragments is not as efficient as ligation of sticky ends; therefore, with cDNA molecules additional procedures are undertaken before ligation with cloning vectors. One approach is to add cDNA small, double stranded molecules with one internal site for a restriction endonuclease; these are termed nucleic acid linkers. Numerous linkers are commercially available with internal restriction for many of the most commonly used restriction enzymes.

Linkers are blunt end ligated to cDNA but since they are added much in excess of the cDNA, the ligation process is reasonably successful. Subsequently the linkers are digested with the appropriate restriction enzyme, which provides the sticky ends for efficient ligation to a vector digested with the same enzyme. This process may be made easier by the addition of adaptors rather than linkers, which are identical except that the sticky ends are performed and so there is no need of restriction digestion following ligation.

Therefore, lastly Linkers are added to the DNA duplexes rendered from the mRNA templates, and the cDNA is cloned into a suitable vector. Once a cDNA derived from a particular gene has been identified, the cDNA becomes an effective probe for screening genomic libraries for isolation of the gene itself.

Because different cell types in eukaryotic organisms express selected subsets of genes, RNA preparations from cells or tissues in which genes of interest are selectively transcribed are enriched for the desired mRNAs. cDNA libraries prepared from such mRNA are representative of the pattern and extent of gene expression that uniquely define particular kinds of differentiated cells.

cDNA libraries of many normal and diseased human cell types are commercially available, including cDNA libraries of many tumour cells. Comparison of normal and abnormal cDNA libraries, in conjunction with two dimensional gel electrophoretic analysis of the proteins produced in normal and abnormal cells is a promising new strategy in clinical medicine to understand disease mechanisms.

Whole Genome Sequencing

Whole genome sequencing is the act of deducing the complete nucleic acid sequence of the genetic code, or genome, of an organism or organelle (specifically, the mitochondrion or chloroplast). The first whole genome sequencing efforts, carried out in 1976 and 1977, focused respectively on the bacteriophages (bacteria-infecting viruses) MS2 and ΦX174, which have relatively small genomes. Since then there have been numerous innovations in the field of DNA sequencing that have expanded the capabilities of the technology. Those innovations, combined with increasing cost-effectiveness in the early 21st century, enabled the routine use of whole genome sequencing in laboratories worldwide, which effectively ushered in a new era of biological discovery. The power of the approach has been realized in the study of human populations and human diseases such as cancer, as well as in the elucidation of whole genome sequences of crop plants, livestock, and other species of scientific or agricultural significance. Thus, it is acknowledged generally that there exists great value in a detailed understanding of the nucleic acid sequence—especially the variations in the sequence that correlate with predisposition to health or disease states or with other properties of societal or economic significance in microbial, animal, and plant populations.

Sequencing Methods: From Genes to Genomes

In 1944 Canadian-born American bacteriologist Oswald Avery and colleagues recognized that the hereditary material passed from parent to offspring was DNA. Subsequent genetic analyses carried out by other scientists on viruses, bacteria, yeast, fruit flies, and nematodes demonstrated that the intentional induction of mutations that disrupted the genetic code, combined with the analysis of observable traits (phenotypes) produced by such mutations, were important approaches to the study of gene function. Such studies, however, were able to query only a fraction of genes in a genome.

The first sequencing methods (the Maxam-Gilbert and Sanger methods), developed in the 1970s, were deployed to reveal the nucleic acid composition of individual genes and the relatively small genomes of certain viruses. The sequencing of larger genomes remained out of reach conceptually, because of high costs and the effort required, until the launch of the Human Genome Project (HGP) in 1990 in the United States. Although the project was not universally embraced, some recognized that technology had evolved to the point where whole genome sequencing of larger genomes could be considered realistically. Particularly important was the development of automated sequencing machines that employed fluorescence instead of radioactive decay for the detection of the sequencing reaction products. Automation offered new possibilities for scaling up the production of DNA sequencing to tackle large genomes.

An early aim of the HGP was to obtain the whole genome sequences of important experimental model organisms, such as the yeast Saccharomyces cerevisiae, the fruit fly Drosophila melanogaster, and the nematode Caenorhabditis elegans. In sequencing those smaller and therefore more-tractable genomes, three outcomes were anticipated. First, the sequences would be of value to the research community, serving to accelerate efforts to understand gene function by using model systems. Second, the experience gained would inform approaches to sequencing the human genome and other similarly sized genomes. Third, functional relationships between sequences of different organisms would be revealed as a consequence of cross-species sequence similarity. Ultimately, with the involvement of more than one thousand scientists worldwide, two human genome sequences. With this development came established methods and analytic standards that were used to sequence other large genomes.

A major challenge for de novo sequencing, in which sequences are assembled for the very first time (such as with the HGP), is the production of individual DNA reads that are of sufficient length and quality to span common repetitive elements, which are a general property of complex genome sequences and a source of ambiguity for sequence assembly. In many of the early de novo whole genome sequencing projects, emphasis was placed on the production of so-called reference sequences, which were of enduring high quality and would serve as the foundation for future experimentation.

An important approach used by many projects that sequenced large genomes involved hierarchical shotgun sequencing, in which segments of genomic DNA were cloned (copied) and arranged into ordered arrays. Those ordered arrays were known as physical maps, and they served to break large genomes into thousands of short DNA fragments. Those short fragments were then aligned, such that identical sequences overlapped, thereby enabling the fragments to be linked together to yield the full-length genomic sequence. The fragments were relatively easy to manipulate in the laboratory, could be apportioned among collaborating laboratories, and were amenable to the detailed error-correction exercises important in generating the high-quality reference sequences sought by HGP scientists. Some genome projects were conducted without the use of such maps, using instead an approach called whole genome shotgun sequencing. This approach avoided the time and expense needed to create physical maps and provided more-rapid access to the DNA sequence.

Whether using physical maps or the whole genome shotgun sequencing approach, the sequencing exercise involved randomly fragmenting either cloned (copied) or native genomic DNA into very short segments that could then be inserted into bacterial cells as plasmids for amplification, producing many copies of the segments, prior to nucleic acid purification and sequence analysis. In a process known as assembly, computer programs were then used to stitch the sequences back together to reconstruct the original DNA sequencing target. Assembly of whole genome shotgun sequencing data was difficult and required sophisticated computer programs and powerful supercomputers, and, even in the years following the completion of the HGP, whole genome shotgun sequence assembly remained a significant challenge for whole genome sequencing projects.

Next-generation Technologies

Although the first whole genome sequences were in themselves technological and scientific feats of significance, the scientific opportunities and the host of technologies those projects spawned have had even greater impacts. Among the most significant technological developments has been in the area of next-generation DNA sequencing technologies for human genome analysis. Certain of those technologies originally were designed to re-sequence genomes (as opposed to de novo sequencing). In re-sequencing, short sequences are produced and aligned computationally to existing reference genome sequences generated, at least initially, using the older de novo sequencing methods. Next-generation sequencing approaches are characterized generally by the massively parallel production of short sequences, in which multiple DNA fragments are generated simultaneously and in sufficient quantity to redundantly represent every base in the target genome.

Gene Therapy

Gene therapy is an experimental technique that uses genes to treat or prevent disease The new DNA usually contains a functioning gene to correct the effects of a disease-causing mutation.

- Gene therapy uses sections of DNA? (usually genes?) to treat or prevent disease.

- The DNA is carefully selected to correct the effect of a mutated gene that is causing disease.

- The technique was first developed in 1972 but has, so far, had limited success in treating human diseases.

- Gene therapy may be a promising treatment option for some genetic diseases?, including muscular dystrophy? and cystic fibrosis?.

- There are two different types of gene therapy depending on which types of cells are treated:

  o Somatic gene therapy: transfer of a section of DNA to any cell of the body that doesn't produce sperm or eggs. Effects of gene therapy will not be passed onto the patient's children.

  o Germline gene therapy: transfer of a section of DNA to cells that produce eggs or sperm. Effects of gene therapy will be passed onto the patient's children and subsequent generations.

Gene Therapy Techniques

Gene Augmentation Therapy

- This is used to treat diseases caused by a mutation that stops a gene from producing a functioning product, such as a protein?.

- This therapy adds DNA containing a functional version of the lost gene back into the cell.

- The new gene produces a functioning product at sufficient levels to replace the protein that was originally missing.

- This is only successful if the effects of the disease are reversible or have not resulted in lasting damage to the body.

- For example, this can be used to treat loss of function disorders such as cystic fibrosis by introducing a functional copy of the gene to correct the disease.

- Gene augmentation therapy.

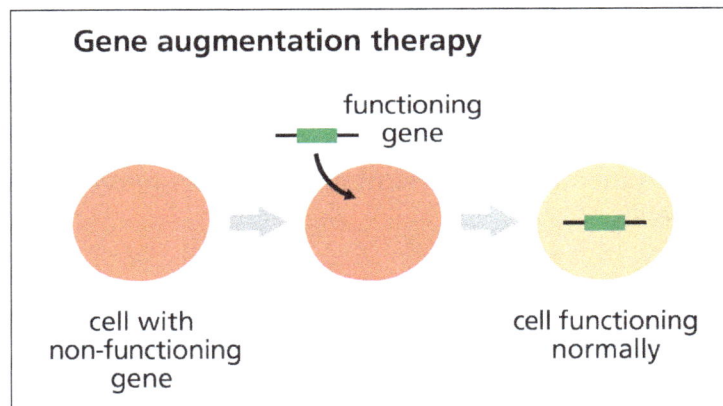

Gene Inhibition Therapy

- Suitable for the treatment of infectious diseases, cancer and inherited disease caused by inappropriate gene activity.

- The aim is to introduce a gene whose product either:

  o inhibits the expression of another gene

  o interferes with the activity of the product of another gene.

- The basis of this therapy is to eliminate the activity of a gene that encourages the growth of disease-related cells.

- For example, cancer is sometimes the result of the over-activation of an oncogene? (gene which stimulates cell growth). So, by eliminating the activity of that oncogene through gene inhibition therapy, it is possible to prevent further cell growth and stop the cancer in its tracks.

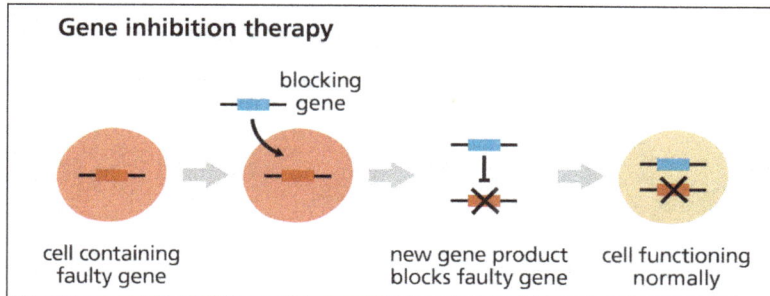

Killing of Specific Cells

- Suitable for diseases such as cancer that can be treated by destroying certain groups of cells.

- The aim is to insert DNA into a diseased cell that causes that cell to die.

- This can be achieved in one of two ways:

  o The inserted DNA contains a "suicide" gene that produces a highly toxic product which kills the diseased cell.

  o The inserted DNA causes expression of a protein that marks the cells so that the diseased cells are attacked by the body's natural immune system.

- It is essential with this method that the inserted DNA is targeted appropriately to avoid the death of cells that are functioning normally.

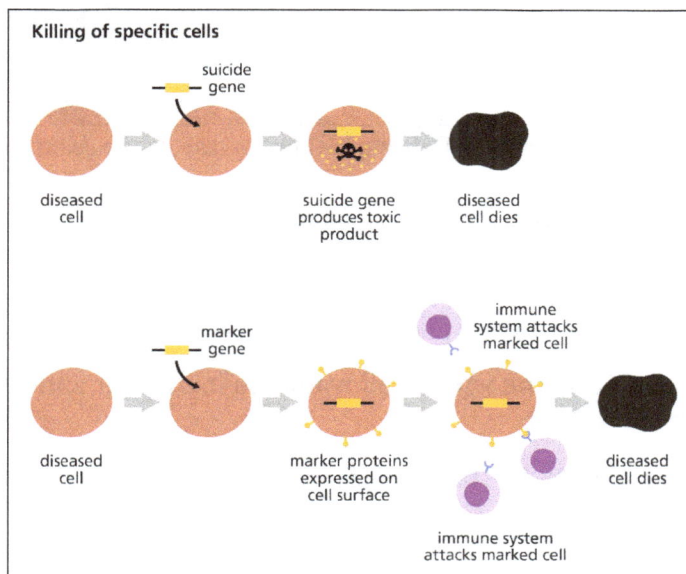

DNA Transfer

- A section of DNA/gene containing instructions for making a useful protein is packaged within a vector, usually a virus?, bacterium? or plasmid?

- The vector acts as a vehicle to carry the new DNA into the cells of a patient with a genetic disease.

- Once inside the cells of the patient, the DNA/gene is expressed by the cell's normal machinery leading to production of the therapeutic protein and treatment of the patient's disease.

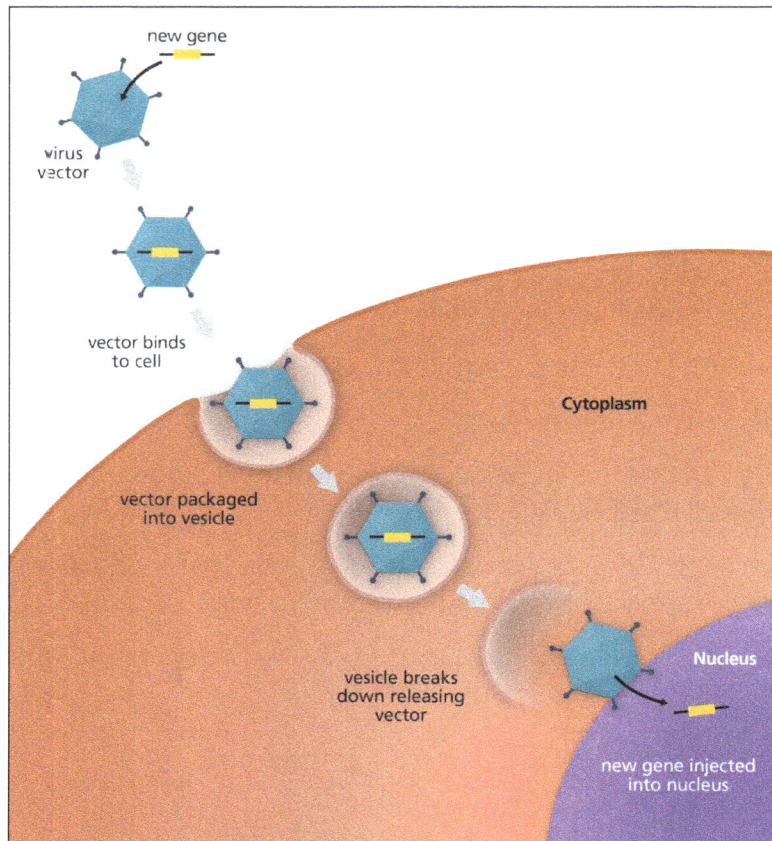

An illustration to show the transfer of a new gene into the nucleus of a cell via a viral vector.

Challenges of Gene Therapy

- Delivering the gene to the right place and switching it on:

  - It is crucial that the new gene reaches the right cell

  - Delivering a gene into the wrong cell would be inefficient and could also cause health problems for the patient

  - Even once the right cell has been targeted the gene has to be turned on

  - Cells sometimes obstruct this process by shutting down genes that are showing unusual activity.

- Avoiding the immune response.

- The role of the immune system is to fight off intruders.

- Sometimes new genes introduced by gene therapy are considered potentially-harmful intruders.

- This can spark an immune response in the patient, that could be harmful to them.

- Scientists therefore have the challenge of finding a way to deliver genes without the immune system 'noticing'.

- This is usually by using vectors that are less likely to trigger an immune response.

- Making sure the new gene doesn't disrupt the function of other genes:

  o Ideally, a new gene introduced by gene therapy will integrate itself into the genome of the patient and continue working for the rest of their lives.

  o There is a risk that the new gene will insert itself into the path of another gene, disrupting its activity.

  o This could have damaging effects, for example, if it interferes with an important gene involved in regulating cell division, it could result in cancer.

- The cost of gene therapy:

  o Many genetic disorders that can be targeted with gene therapy are extremely rare.

  o Gene therapy therefore often requires an individual, case-by-case approach. This may be effective, but may also be very expensive.

References

- Uoh-genetic, view, lessons: teachengineering.org, Retrieved, 24 March, 2019

- Genetically-modified-organism, science: britannica.com, Retrieved, 17 February, 2019

- Dna-gene-libraries-construction-genomic-libraries-and-cdna-libraries, gene-libraries, gene: biologydiscussion.com, Retrieved, 28 April, 2019

- Whole-genome-sequencing, science: britannica.com, Retrieved, 21 July, 2019

- What-is-gene-therapy, facts: yourgenome.org, Retrieved, 15 January, 2019

- Human-Genome-Project, event: britannica.com, Retrieved 18 May, 2019

Agricultural Biotechnology

The branch of agricultural science which makes use of scientific techniques and tools to modify living organisms such as plants, animals and microorganisms is called agricultural biotechnology. Genetic engineering, tissue culture and molecular markers are some of the techniques used under this domain. All the diverse applications of agricultural biotechnology for the purpose of plant breeding and gene transfer in plants have been carefully analyzed in this chapter.

Agricultural biotechnology is the use of different scientific techniques to modify plants and animals. The undesirable characteristics like susceptibility to diseases and low productivity are bred out. If there is a particular trait that the plant or animal can benefit from, it can be bred in by using a gene that contains the characteristic.

Biotechnology has especially been beneficial in improving agricultural productivity and increasing the resistance of plants to diseases. Scientists do this by studying the DNA. They first identify the gene that would be beneficial to the plant or animal then work with the characteristics conferred in a precise and exact manner to achieve the desired outcome.

Advantages of Agricultural Biotechnology

Biotechnology has been beneficial in many ways. First, stabilized plants that have higher yields have been produced successfully. The resistance of these plants to pests, diseases and abiotic factors such as rainfall has played a major role in increasing the yields.

Animal feeds are being improved by biotechnology to increase their nutrient intake and reduce environmental wastes.

Another advantage of biotechnology is that it has led to the development of better vaccines that don't necessarily have to be stored in very cold temperatures. Penicillin, one of the most important components of antibiotics was produced through biotechnology.

Uses of Agricultural Biotechnology

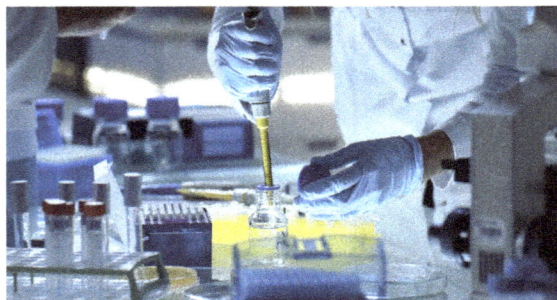

1. Genetic Engineering: Genetic engineering, also referred to as genetic improvement or modification,

is the movement of a gene from one organism to another. This process allows for the transfer of a useful characteristic into an organism by inserting it with a gene containing the particular trait. In crops, genetic engineering has been used to increase productivity and resistance to weeds and harsh weather conditions.

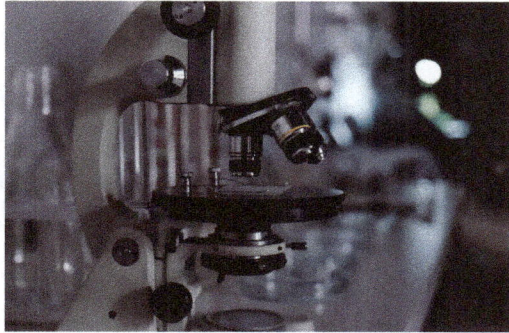

2. Molecular Markers: Breeding was previously concerned with the removal and insertion of desirable physical traits, an example being the aggression of Bulldogs. By studying the DNA, scientists were able to find molecular markers that showed traits that were not visible. Using molecular markers, breeding has been made more precise and accurate, and this has countered the undesirable characteristics that may have appeared in future generations.

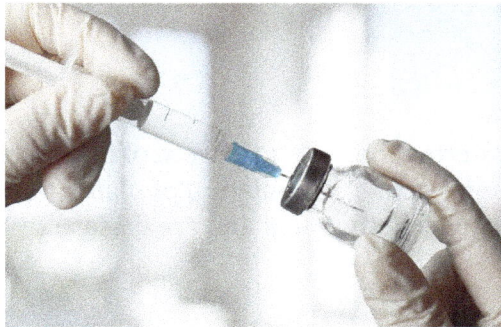

3. Vaccines: Biotechnology is used for making vaccines for both animals and human beings. These vaccines are better than the traditional ones because they are cheaper, safer, and can survive warmer tropical temperatures. Vaccines to prevent new infections have also been developed using biotechnology.

4. Genomics: A genome is an entire set of chromosomes found in the DNA, and through the study of genomes and genetic mechanisms, breakthroughs have been made in biotechnology. Through

genomics, the structure, function, location, and impact of a particular gene and genome are identi-fied. This makes it easy to determine the characteristics that will be transferred to another organ-ism and the exact results of the transfer of the gene.

5) Tissue Culture: This technique is used to produce a plant that is free from undesirable charac-teristics, which are mostly, diseases. A disease-free plant part is used to generate types that are dis-ease-free. The different types of plants in which tissue culture works include bananas, avocados, mangoes, coffee, and papaya, among others.

Plant Breeding

Plant breeding is a method of altering the genetic pattern of plants to increase their value and util-ity for human welfare. It is a purposeful manipulation of plants to create desired plant types that are better suited for cultivation, give better yield and are disease resistant. Plant breeding is done for the following objectives:

1. Increase the crop yield,

2. Improve the quality of the crop,

3. Increase tolerance to environmental conditions like salinity. extreme temperatures and drought,

4. Develop a resistance to pathogens,

5. Increase tolerance to the insect pest.

Steps for Different Plant Breeding Methods

The main steps of the plant breeding program are as follows:

1. Collection of Variability: Wild varieties species and relatives of the cultivated species having desired traits should be collected and preserved. The entire collection having all the diverse alleles for all genes in a given crop is called germplasm collection. Germplasm conservation can be done following ways:

• In situ conservation – It can be done with the help of forests and Natural Reserves.

- Ex situ conservation- it is done through botanical gardens, seed banks.

2. Evaluation and Selection of Parents: The germplasm collected is evaluated to identify the plants with desirable characters. It is made sure that only the pure lines are selected. The selected plants are multiplied and used in the process of hybridization.

3. Hybridization: The Pollen Grain from one desired parent plant selected as a male parent is collected and dusted over another plant which is considered as the female parent.

4. Selection and Testing of Superior Recombinants: Progeny obtained after crossing are evaluated for the desired combination of characters. These are self-pollinated for several generations till there is a state of uniformity so that the characters will not segregate further.

5. Testing Release and Commercialization of New Cultivars: The selected plants are evaluated by growing the plants in an experimental field and the performance is recorded. This is done for at least 3 growing Seasons at different locations in the country.

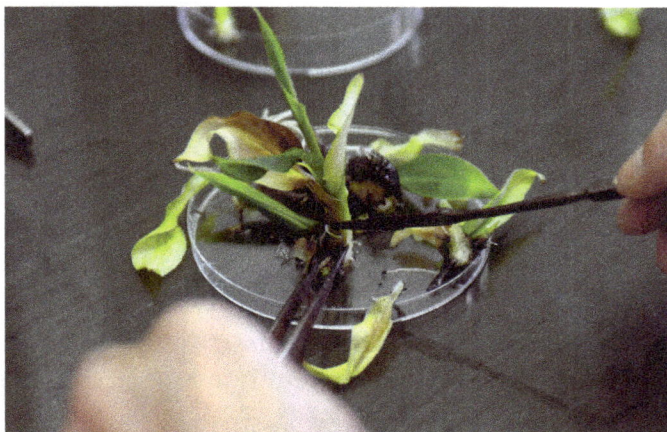

Plant breeding

Plant Breeding for Disease Resistance

Some of the diseases caused by plants are:

Fungi	Brown rust of wheat
	Red rot of sugarcane
	Late blight of potato
Bacteria	Black rot of crucifers
Viruses	Tobacco mosaic turnip music

The basic objective of breeding for disease resistance is to develop inherent quality in the plant to prevent the pathogen from causing the disease. Such varieties of plants are called resistant plants. The basic method is the same as normal hybridization. For hybridization resistant plant should be available for breeding.

Some diseases resistant plants developed are:

Crop	Variety	Resistant to Disease
Wheat	Himgiri	Leaf & stipe rust
Brassica	Pusa Swarnim	White rust
Cauliflower	Pusa Shubra	Black rot and curl blight
Chilli	Pusa Sadabahar	Chilli mosaic virus

Mutation Breeding

If resistant variety is not available, the resistance can be developed by inducing mutations in the plant through diverse means and then by screening the plant material for resistance.

Mutation is changed in the base sequence of the genes, induced by certain Chemicals or radiations. The resistant plants developed can be multiplied directly or can be used in other breeding experiments.

Plant Breeding for Developing a Resistance to Insect Pest

Resistance can be developed by following ways:

- Development of morphological characters like hairy leaves in cotton and wheat develop vector resistance from jassids beetle.

- Solid stem in wheat lead to resistance from stem borers.

- Biochemical characters provide resistance to insects and pests. For example, the high aspartic acid and low nitrogen and sugar content in maize leads to resistance to maize stem borers.

- Smooth leaves and nectarless cotton develop resistance from bollworms.

Some pest-resistant varieties are:

Crop	Variety	Insect pest
Brassica	Pusa Gaurav	Aphids
Flat bean	Pusa sem	Jassids, aphids
Okra	Pusa Sawni	Shoot and fruit bores

Plant Breeding for Improved Food Quality

Biofortification is a method in which crops are bred for higher levels of vitamins, minerals, and fats. Due to this problem of malnutrition can be overcome. Following objectives were considered for the breeding program:

- Protein content and quality,

- Oil content and quality,

- Vitamin content,

- Micronutrient content and quality.

Some examples of biofortification:

- Fortified Maize having twice the amount of amino acids lysine and tryptophan.
- Atlas 66 wheat has a high protein content.
- Iron-fortified rice having 5 times more iron.
- Vegetable crops like carrot and spinach have more vitamins and minerals.
- Vitamin C enriched bitter gourd.

Gene Transfer in Plants

The gene transfer techniques in plant genetic transformation are broadly grouped into two categories:

- Vector-mediated gene transfer
- Direct or vector less DNA transfer

The salient features of the commonly used gene (DNA) transfer methods are given in table.

Table: Gene transfer (DNA delivery) methods in plants.

Method	Salient Features
I. Vector-mediated gene transfer	
Agrobacterium(Ti plasmid).mediated gene transfer plant viral vectors	Very efficient, but limited to a selected group of plants Ineffective method, hence not widely used.
II. Direct or vectorless DNA transfer	
(A) Physical methods	
Electroporation	Mostly confined to protoplasts that can be regenerated viable plants. Many cereal crops developed.
Microprohectile (particle bombardment)	Very successful method used for a wide range of plants/tissues. Risk of gene rearrangement high.
Microinjection	Limite use since only one cell can be microinjected at a time. Technical personnel should be highly skilled.
Liposome fusion	Confined to protoplasts that can be regenerated into viable whole plants.
Silicon carbide fibres	Requires regenerable cell suspensions. The fibres, However, require careful handling.
(B) Chemical methods	
Polyethylene glycol (PEG)-mediated	Confined to protoplasts. Regeneration of fertile plants is frequently problematical.
Diethylaminoehtyl (DEAE) Dextran-mendiated	Does not result in stable transformants.

Method I: Vector-mediated Gene Transfer

Vector-mediated gene transfer is carried out either by Agrobacterium-mediated transformation or by use of plant viruses as vectors.

Agrobacterium-mediated Gene Transfer

Agrobacterium tumefaciens is a soil-borne, Gram-negative bacterium. It is rod shaped and motile, and belongs to the bacterial family of Rhizobiaceae. A. tumefaciens is a phytopathogen, and is treated as the nature's most effective plant genetic engineer.

Some workers consider this bacterium as the natural expert of inter-kingdom gene transfer. In fact, the major credit for the development of plant transformation techniques goes to the natural unique capability of A. tumefaciens. Thus, this bacterium is the most beloved by plant biotechnologists.

There are mainly two species of Agrobacterium:

- A. tumefaciens that induces crown gall disease.

- A. rhizogenes that induces hairy root disease.

Crown Gall Disease and Ti Plasmid

Smith and Townsend postulated that a bacterium was the causative agent of crown gall tumors, although its importance was recognized much later. As A. tumefaciens infects wounded or damaged plant tissues, in induces the formation of a plant tumor called crown gall. The entry of the bacterium into the plant tissues is facilitated by the release of certain phenolic compounds (acetosyringone, hydroxyacetosyringone) by the wounded sites.

Formation of a Crown Gall Tumor in a plant infected with Agrobacterium tumefaciens.

Crown gall formation occurs when the bacterium releases its Ti plasmid (tumor- inducing plasmid) into the plant cell cytoplasm. A fragment (segment) of Ti plasmid, referred to as T-DNA, is actually transferred from the bacterium into the host where it gets integrated into the plant cell chromosome (i.e. host genome). Thus, crown gall disease is a naturally evolved genetic engineering process.

The T-DNA carries genes that code for proteins involved in the biosynthesis of growth hormones (auxin and cytokinin) and novel plant metabolites namely opines — amino acid derivatives and agropines — sugar derivatives.

The growth hormones cause plant cells to proliferate and form the gall while opines and agropines are utilized by A. tumefaciens as sources of carbon and energy. As such, opines and agropines are not normally part of the plant metabolism (neither produced nor metabolised). Thus, A. tumefaciens genetically transforms plant cells and creates a biosynthetic machinery to produce nutrients for its own use.

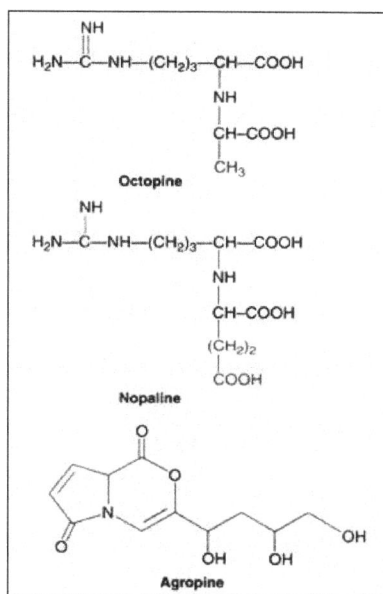

Structures of three Opines-Octopine, Nopaline and Agropine.

As the bacteria multiply and continue infection, grown gall develops which is a visible mass of the accumulated bacteria and plant material. Crown gall formation is the consequence of the transfer, integration and expression of genes of T-DNA (or Ti plasmid) of A. tumefaciens in the infected plant.

The genetic transformation leads to the formation of crown gall tumors, which interfere with the normal growth of the plant. Several dicotyledonous plants (dicots) are affected by crown gall disease e.g. grapes, roses, stone-fruit trees.

Organization of Ti Plasmid

The Ti plasmids (approximate size 200 kb each) exist as independent replicating circular DNA molecules within the Agrobacterium cells. The T-DNA (transferred DNA) is variable in length in the range of 12 to 24 kb, which depends on the bacterial strain from which Ti plasmids come. Nopaline strains of Ti plasmid have one T-DNA with length of 20 kb while octopine strains have two T-DNA regions referred to as TL and TR that are respectively 14 kb and 7 kb in length.

A diagrammatic representation of a Ti plasmid is depicted in. The Ti plasmid has three important regions.

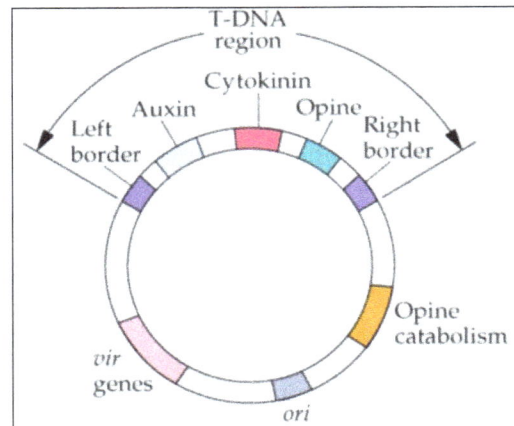

A diagrammatic representation of a Ti Plasmid.

1. T-DNA region: This region has the genes for the biosynthesis of auxin (aux), cytokinin (cyt) and opine (ocs), and is flanked by left and right borders. These three genes-aux, cyto and ocs are referred to as oncogenes, as they are the determinants of the tumor phenotype.

    T-DNA borders — A set of 24 kb sequences present on either side (right and left) of T-DNA are also transferred to the plant cells. It is now clearly established that the right border is more critical for T-DNA transfer and tumori-genesis.

2. Virulence region: The genes responsible for the transfer of T-DNA into the host plant are located outside T-DNA and the region is referred to as vir or virulence region. Vir region codes for proteins involved in T-DNA transfer. At least nine vir-gene operons have been identified. These include vir A, vir G, vir B_1, vir C_1, vir D_1, D_2 and D_4, and vir E_1, and E_2.

3. Opine catabolism region: This region codes for proteins involved in the uptake and metabolisms of opines. Besides the above three, there is ori region that is responsible for the origin of DNA replication which permits the Ti plasmid to be stably maintained in A. tumefaciens.

T-DNA Transfer and Integration

The process of T-DNA transfer and it integration into the host plant genome is depicted in figure, and is briefly described.

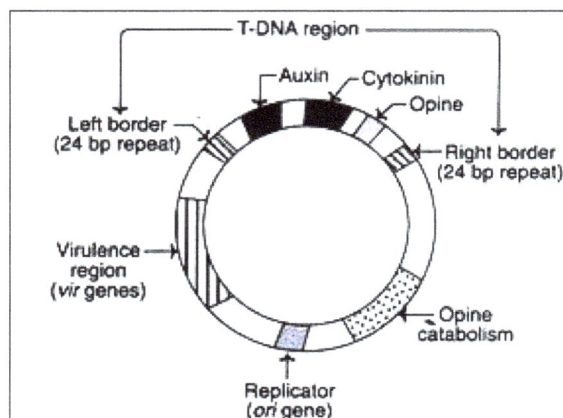

A diagrammatic representation of a Ti Plasmid.

1. Signal induction to Agrobacterium: The wounded plant cells release certain chemicals-phenolic compounds and sugars which are recognized as signals by Agrobacterium. The signals induced result in a sequence of biochemical events in Agrobacterium that ultimately helps in the transfer of T-DNA of T-plasmid.

2. Attachment of Agrobacterium to plant cells: The Agrobacterium attaches to plant cells through polysaccharides, particularly cellulose fibres produced by the bacterium. Several chromosomal virulence (chv) genes responsible for the attachment of bacterial cells to plant cells have been identified.

3. Production of virulence proteins: As the signal induction occurs in the Agrobacterium cells attached to plant cells, a series of events take place that result in the production of virulence proteins. To start with, signal induction by phenolics stimulates vir A which in turn activates (by phosphorylation) vir C. This induces expression of virulence genes of Ti plasmid to produce the corresponding virulence proteins (D_1, D_2, E_2, B etc.). Certain sugars (e.g. glucose, galactose, xylose) that induce virulence genes have been identified.

4. Production of T-DNA strand: The right and left borders of T-DNA are recognized by vir D_1/vir D_2 proteins. These proteins are involved in the production single-stranded T-DNA (ss DNA), its protection and export to plant cells. The ss T-DNA gets attached to vir D2.

5. Transfer of T-DNA out of Agrobacterium: The ss T-DNA — vir D2 complex in association with vir G is exported from the bacterial cell. Vir B products form the transport apparatus.

6. Transfer of T-DNA into plant cells and integration: The T-DNA-vir D2 complex crosses the plant plasma membrane. In the plant cells, T-DNA gets covered with vir E2. This covering protects the T-DNA from degradation by nucleases; vir D2 and vir E2 interact with a variety of plant proteins which influences T-DNA transport and integration.

The T-DNA-vir D_2-vir E_2 — plant protein complex enters the nucleus through nuclear pore complex. Within the nucleus, the T-DNA gets integrated into the plant chromosome through a process referred to illegitimate recombination. This is different from the homologous recombination, as it does not depend on the sequence similarity.

Hairy Root Disease of A. Rhizogenes — R_1 Plasmids

Agrobacterium rhizogenes can also infect plants. But this results in hairy roots and not crown galls as is the case with A. tumefaciens. The plasmids, of A. rhizogenes have been isolated and characterized. These plasmids, referred to as Ri plasmids, (Ri stands for Root inducing) are of different types. Some of the Ri plasmid strains possess genes that are homologous to Ti plasmid e.g. auxin biosynthetic genes.

Instead of virulence genes, Ri plasmids contain a series of open reading frames on the T-DNA. The products of these genes are involved in the metabolism of plant growth regulators which gets sensitized to auxin and leads to root formation.

Vectors of A. Rhizogenes

As it is done with A tumefaciens, vectors can be constructed by using A. rhizogenes. These vectors are alternate strategies for gene transfer. However, employment of A. rhizogene-based vectors for plant transformation is not common since more efficient systems of A. tumefaciens have been developed.

Importance of Hairy Roots

Hairy roots can be cultured in vitro, and thus are important in plant biotechnology. Hairy root systems are useful for the production of secondary metabolites, particularly pharmaceutical proteins.

Ti Plasmid-derived Vector Systems

The ability of Ti plasmid of Agrobacterium to genetically transform plants has been described. It is possible to insert a desired DNA sequence (gene) into the T-DNA region (of Ti plasmid), and then use A. tumefaciens to deliver this gene(s) into the genome of plant cell.

In this process, Ti plasmids serve as natural vectors. However, there are several limitations to use Ti plasmids directly as cloning vectors:

1. Ti plasmids are large in size (200-800 kb). Smaller vectors are preferred for recombinant experiments. For this reason, large segments of DNA of Ti plasmid, not essential for cloning, must be removed.

2. Absence of unique restriction enzyme sites on Ti plasmids.

3. The phytohormones (auxin, cytokinin) produced prevent the plant cells being regenerated into plants. Therefore auxin and cytokinin genes must be removed.

4. Opine production in transformed plant cells lowers the plant yield. Therefore opine synthesizing genes which are of no use to plants should be removed.

5. Ti plasmids cannot replicate in E. coli. This limits their utility as E. coli is widely used in recombinant experiments. An alternate arrangement is to add an origin of replication to Ti plasmid that allows the plasmid to replicate in E. coli.

Considering the above limitations, Ti plasmid-based vectors with suitable modifications have been constructed.

These vectors are mainly composed of the following components:

1. The right border sequence of T-DNA which is absolutely required for T-DNA integration into plant cell DNA.

2. A multiple cloning site (poly-linker DNA) that promotes the insertion of cloned gene into the region between T-DNA borders.

3. An origin of DNA replication that allows the plasmids to multiply in E. coli.

4. A selectable marker gene (e.g. neomycin phosphotransferase) for appropriate selection of the transformed cells.

Two types of Ti plasmid-derived vectors are used for genetic transformation of plants— co-integrate vectors and binary vectors.

Co-integrate Vector

In the co-integrate vector system, the disarmed and modified Ti plasmid combines with an intermediate cloning vector to produce a recombinant Ti plasmid.

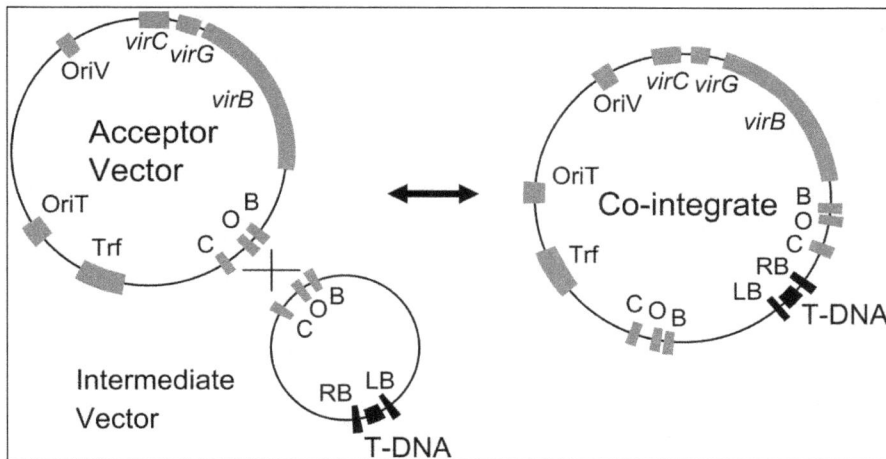

Figure indicates, Cointegrate Vector System (vir- Ti plasmid virulence region; pBR322-Bacterial plasmid 322; LB-Left border; RB-Right ; MCS-Multiple cloning site; PTM-Plant transformation marker; RES- Bacterial resistance Marker col E,-Origin of a replication from col E, Plasmid; ori T-Origin of transfer site for conjugative plasmid mobilization).

Production of Disarmed Ti Plasmid

The T-DNA genes for hormone biosynthesis are removed (disarmed). In place of the deleted DNA, a bacterial plasmid (pBR322) DNA sequence is incorporated. This disarmed plasmid, also referred to as receptor plasmid, has the basic structure of T-DNA (right and left borders, virulence genes etc.) necessary to transfer the plant cells.

Construction of Intermediate Vector

The intermediate vector is constructed with the following components:

i.   A pBR322 sequence DNA homologous to that found in the receptor Ti plasmid.

ii. A plant transformation marker (PTM) e.g. a gene coding for neomycin phosphotransferase II (npt II). This gene confers resistance to kanamycin in the plant cells and thus permits their isolation.

iii. A bacterial resistance marker e.g. a gene coding for spectinomycin resistance. This gene confers spectinomycin resistance to recipient bacterial cells and thus permits their selective isolation.

iv. A multiple cloning site (MCS) where foreign genes can be inserted.

v. A Co/E1 origin of replication which allows the replication of plasmid in E. coli but not in Agrobacterium.

vi. An oriT sequence with basis of mobilization (bom) site for the transfer of intermediate vector from E. coli to Agrobacterium.

Production and use of Co-integrate Vectors

The desired foreign gene (target-gene) is first cloned in the multiple cloning site of the intermediate vector. The cloning process is carried out in E. coli, the bacterium where the cloning is most efficient. The intermediate vector is mated with Agrobacterium so that the foreign gene is mobilised into the latter.

The transformed Agrobacterium cells with receptor Ti plasmid and intermediate vector are selectively isolated when grown on a minimal medium containing spectinomycin. The selection process becomes easy since E. coli does not grow on a minimal medium in which Agrobacterium grows.

Within the Agrobacterium cells, intermediate plasmid gets integrated into the receptor Ti plasmid to produce co-integrate plasmid. This plasmid containing plant transformation marker (e.g. npt II) gene and cloned target gene between T-DNA borders is transferred to plant cells. The transformed plant cells can be selected on a medium containing kanamycin when the plant and Agrobacterium cells are incubated together.

Advantages of Co-integrate Vector

i.   Target genes can be easily cloned.

ii. The plasmid is relatively small with a number of restriction sites.

iii. Intermediate plasmid is conveniently cloned in E. coli and transferred to Agrobacterium.

Binary Vector

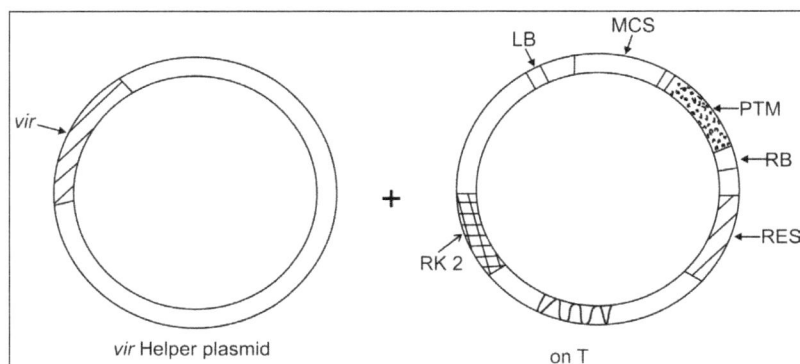

Binary Vector System (vir- Ti plasmid virulence region; LB-Left border; RB-Right border; MCS-Multiple cloning site; PTM-Plant transformation marker; RES-Bacterial resistande marker; oriT-origin of transfer site for conjugative plasmid mobilization; RK$_2$-origin of replication from plasmid).

The binary vector system consists of an Agrobacterium strain along with a disarmed Ti plasmid called vir helper plasmid (the entire T-DNA region including borders deleted while vir gene is retained). It may be noted that both of them are not physically linked (or integrated). A binary vector with T-DNA can replicate in E. coli and Agrobacterium.

A diagrammatic representation of a typical binary vector system is depicted in figure. The binary vector has the following components.

- Left and right borders that delimit the T-DNA region.

- A plant transformation marker (PTM) e.g. npt II that confers kanamycin resistance in plant transformed cells.

- A multiple cloning site (MCS) for introducing target/foreign genes.

- A bacterial resistance marker e.g. tetracycline resistance gene for selecting binary vector colonies in E. coli and Agrobacterium.

- oriT sequence for conjugal mobilization of the binary vector from E. coli to Agrobacterium.

- A broad host-range origin of replication such as RK2 that allows the replication of binary vector in Agrobacterium.

Production and use of Binary Vector

The target (foreign) gene of interest is inserted into the multiple cloning site of the binary vector. In this way, the- target gene is placed between the right and left border repeats and cloned in E. coli. By a mating process, the binary vector is mobilised from E. coli to Agrobacterium. Now, the virulence gene proteins of T-DNA facilitate the transfer of T-DNA of the vector into plant cells.

Advantages of Binary Vector

- The binary vector system involves only the transfer of a binary plasmid to Agrobacterium without any integration. This is in contrast to co-integrate vector system wherein the intermediate vector is transferred and integrated with disarmed Ti plasmid.

- Due to convenience, binary vectors are more frequently used than co-integrate vectors.

Plant Transformation Technique using Agrobacterium

Agrobacterium-mediated technique is the most widely used for the transformation of plants and generation of transgenic plants. The important requirements for gene transfer in higher plants through Agrobacterium mediation are listed.

- The explants of the plant must produce phenolic compounds (e.g. autosyringone) for activation of virulence genes.

- Transformed cells/tissues should be capable to regenerate into whole plants.

In general, most of the Agrobacterium-mediated plant transformations have the following basic protocol.

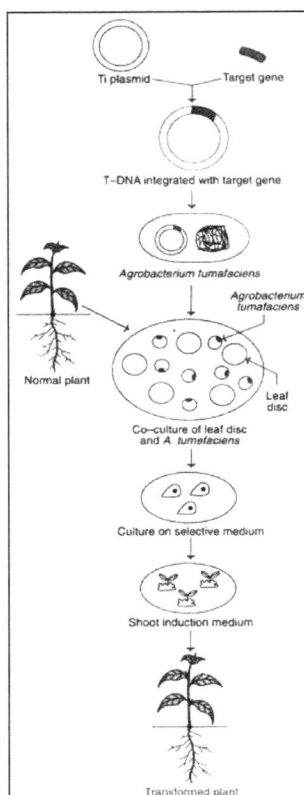

Transformation Technique using Agrobacterium- mediated gene transfer.

i.   Development of Agrobacterium carrying the co-integrate or binary vector with the desired gene.

ii. Identification of a suitable explant e.g. cells, protoplasts, tissues, calluses, organs.

iii. Co-culture of explants with Agrobacterium.

iv. Killing of Agrobacterium with a suitable antibiotic without harming the plant tissue.

v.   Selection of transformed plant cells.

vi. Regeneration of whole plants.

Advantages of Agrobacterium-mediated Transformation

i.   This is a natural method of gene transfer.

ii. Agrobacterium can conveniently infect any explant (cells/tissues/organs).

iii. Even large fragments of DNA can be efficiently transferred.

iv. Stability of transferred DNA is reasonably good.

v.   Transformed plants can be regenerated effectively.

Limitations of Agrobacterium-mediated Transformation

i.   There is a limitation of host plants for Agrobacterium, since many crop plants (monocotyledons e.g. cereals) are not infected by it. In recent years, virulent strains of Agrobacterium that can infect a wide range of plants have been developed.

ii. The cells that regenerate more efficiently are often difficult to transform, e.g. embryonic cells lie in deep layers which are not easy targets for Agrobacterium.

Virus-mediated Gene Transfer (Plant Viruses as Vectors)

Plant viruses are considered as efficient gene transfer agents as they can infect the intact plants and amplify the transferred genes through viral genome replication. Viruses are natural vectors for genetic engineering. They can introduce the desirable gene(s) into almost all the plant cells since the viral infections are mostly systemic.

Plant Viruses are Non-integrative Vectors

The plant viruses do not integrate into the host genome in contrast to the vectors based on T-DNA of A. tumefaciens which are integrative. The viral genomes are suitably modified by introducing desired foreign genes. These recombinant viruses are transferred, multiplied and expressed in plant cells. They spread systemically within the host plant where the new genetic material is expressed.

Criteria for a Plant Virus Vector

An ideal plant virus for its effective use in gene transfer is expected to posses the following characteristics:

i. The virus must be capable of spreading from cell to cell through plasmodesmata.

ii. The viral genome should be able to replicate in the absence of viral coat protein and spread from cell to cell. This is desirable since the insertion of foreign DNA will make the viral genome too big to be packed.

iii. The recombinant viral genome must elicit little or no disease symptoms in the infected plants.

iv. The virus should have a broad host range.

v. The virus with DNA genome is preferred since the genetic manipulations involve plant DNA.

The three groups of viruses — caulimoviruses, Gemini viruses and RNA viruses that are used as vectors for gene transfer in plants are briefly described.

Caulimoviruses as Vectors

The caulimoviruses contain circular double- stranded DNA, and are spherical in shape. Caulimoviruses are widely distributed and are responsible for a number of economically important diseases in various crops. The caulimovirus group has around 15 viruses and among these cauliflower mosaic virus (CaMV) is the most important for gene transfer. The other caulimoviruses include carnation etched virus, dahlia mosaic virus, mirabilis mosaic virus and strawberry vein banding virus.

Cauliflower Mosaic Virus (CaMV)

CaMV infects many plants (e.g. members of Cruciferae, Datura) and can be easily transmitted, even mechanically. Another attractive feature of CaMV is that the infection is systemic, and large quantities of viruses are found in infected cells.

A diagrammatic view of the CaMV genetic map is depicted. The genome of CaMV consists of a 8 kb (8024 bp) relaxed but tightly packed circular DNA with six major and two minor coding regions. The genes II and VII are not essential for viral infection.

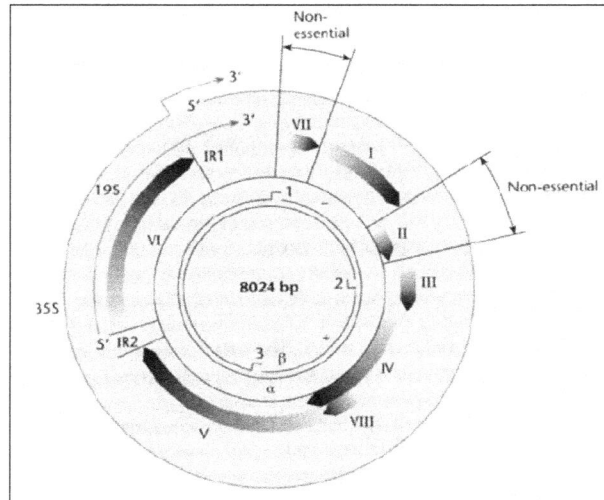

A diagrammatic representation of the Genetic map of Cauliflower Mosaic Virus Genome
(I...VIII represent coding regions; IR 1and IR 2 are intergeneric regions; The outside dotted circle represents 30S transcript; the two circular lines at the centre indicate viral DNA strands).

Use of CaMV in Gene Transfer

For appropriate transmission of CaMV, the foreign DNA must be encapsulated in viral protein. Further, the newly inserted foreign DNA must not interfere with the native assembly of the virus. CaMV genome does not contain any non-coding regions wherein foreign DNA can be inserted. It is fortunate that two genes namely gene II and gene VII have no essential functions for the virus. It is therefore possible to replace one of them and insert the desired foreign gene.

Gene II of CaMV has been successfully replaced with a bacterial gene encoding dihydrofolate reductase that provides resistance to methotrexate. When the chimeric CaMV was transmitted to turnip plants, they were systemically infected and the plants developed resistance to methotrexate.

Limitations of CaMV as a Vector

i.   CaMV vector has a limited capacity for insertion of foreign genes.

ii. Infective capacity of CaMV is lost if more than a few hundred nucleotides are introduced.

iii. Helper viruses cannot be used since the foreign DNA gets expelled and wild-type viruses are produced.

Gemini Viruses as Vectors

The Gemini viruses are so named because they have geminate (Gemini literally means heavenly twins) morphological particles i.e. twin and paired capsid structures. These viruses are characterized by possessing one or two single-stranded circular DNAs (ss DNA). On replications, ss DNA forms an intermediate double-stranded DNA.

The Gemini viruses can infect a wide range of crop plants (monocotyledons and dicotyledons) which attract plant biotechnologists to employ these viruses for gene transfer. Curly top virus (CTV) and maize streak virus (MSV) and bean golden mosaic virus (BGMV) are among the important Gemini viruses.

It has been observed that a large number of replicative forms of a Gemini virus genome accumulate inside the nuclei of infected cells. The single-stranded genomic DNA replicates in the nucleus to form a double-stranded intermediate.

Gemini virus vectors can be used to deliver, amplify and express foreign genes in several plants/ explants (protoplasts, cultured cells). However, the serious drawback in employing Gemini viruses as vectors is that it is very difficult to introduce purified viral DNA into the plants. An alternate arrangement is to take the help of Agrobacterium and carry out gene transfer.

RNA Plant Viruses as Vectors

There are mainly two type's single-stranded RNA viruses:

1. Mono-partite viruses: These viruses are usually large and contain undivided genomes for all the genetic information e.g. tobacco mosaic virus (TMV).

2. Multipartite viruses: The genome in these viruses is divided into small RNAs which may be in the same particle or different particles, e.g. brome mosaic virus (BMV). HMV contains four RNAs divided between three particles. Plant RNA viruses, in general, are characterized by high level of gene expression, good efficiency to infect cells and spread to different tissues. But the major limitation to use them as vectors is the difficulty of joining RNA molecules in vitro.

Use of cDNA for Gene Transfer

Complementary DNA (cDNA) copies of RNA viruses are prepared in vitro. The cDNA so generated can be used as a vector for gene transfer in plants. This approach is tedious and cumbersome. However, some success has been reported. A gene sequence encoding chloramphenicol resistance (enzyme- chloramphenicol acetyltransferase) has been inserted into brome mosaic virus genome. This gene expression, however, has been confined to protoplasts.

Limitations of Viral Vectors in Gene Transfer

The ultimate objective of gene transfer is to transmit the desired genes to subsequent generations. With virus vectors, this is not possible unless the virus is seed-transmitted. However, in case of vegetatively propagated plants, transmission of desired traits can be done e.g. potatoes. Even in these plants, there is always a risk for the transferred gene to be lost anytime. For the reasons referred above, plant biotechnologists prefer to insert the desired genes of interest into a plant chromosome.

Method II: Direct or Vector-less DNA Transfer

The term direct or vector less transfer of DNA is used when the foreign DNA is directly introduced into the plant genome. Direct DNA transfer methods rely on the delivery of naked DNA into the plant cells. This is in contrast to the Agrobacterium or vector-mediated DNA transfer which may be regarded as indirect methods. Majority of the direct DNA transfer methods are simple and effective. And in fact, several transgenic plants have been developed by this approach.

Limitations of Direct DNA transfer

The major disadvantage of direct gene transfer is that the frequency of transgene rearrangements is high. This results in higher transgene copy number, and high frequencies of gene silencing.

Types of Direct DNA Transfer

The direct DNA transfer can be broadly divided into three categories.

1. Physical gene transfer methods—electro- portion, particle bombardment, microinjection, liposome fusion, silicon carbide fibres.

2. Chemical gene transfer methods—Polyethylene glycol (PEG)-mediated, diethyl amino ethyl (DEAE) dextran-mediated, calcium phosphate precipitation.

3. DNA imbibition by cells/tissues/organs.

Physical Gene Transfer Methods

An overview of the general scheme for the production of transgenic plants by employing physical transfer methods is depicted. Some details of the different techniques are described.

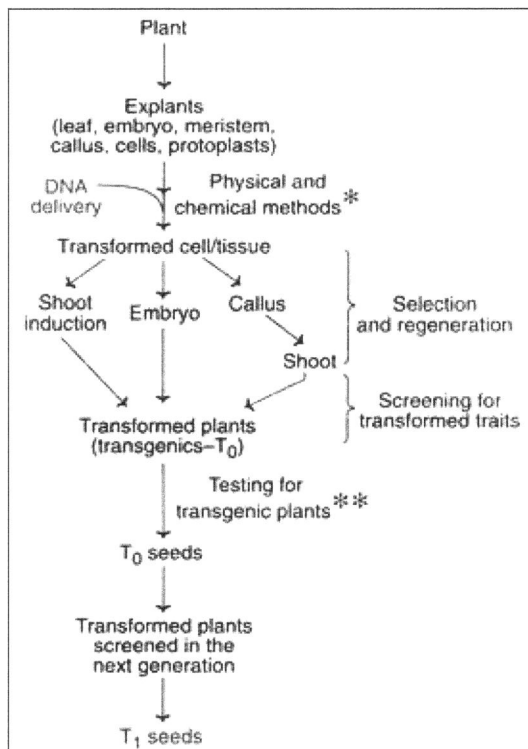

An overview of the protocol for The Protocol for the Production of Transgenic Plants using direct DNA delivery methods (*Electroporation, microinjection, macroinjection, bombardment, etc. ** polymerase chain reaction, Southem hybridization).

1. Electroporation:

Electroporation basically involves the use of high field strength electrical impulses to reversibly permeabilize the cell membranes for the uptake of DNA. This technique can be used for the delivery of DNA into intact plant cells and protoplasts.

The plant material is incubated in a buffer solution containing the desired foreign/target DNA, and subjected to high voltage electrical impulses. This results in the formation of pores in the plasma membrane through which DNA enters and gets integrated into the host cell genome.

In the early years, only protoplasts were used for gene transfer by electroporation. Now a days, intact cells, callus cultures and immature embryos can be used with suitable pre- and post-electroporation treatments. Electroporation has been successfully used for the production of transgenic plants of many cereals e.g. rice, wheat, maize.

Advantages of electroporation:

i.   This technique is simple, convenient and rapid, besides being cost-effective.

ii. The transformed cells are at the same physiological state after electroporation.

iii. Efficiency of transformation can be improved by optimising the electrical field strength, and addition of spermidine.

Limitations of electroporation:

i.   Under normal conditions, the amount of DNA delivered into plant cells is very low.

ii. Efficiency of electroporation is highly variable depending on the plant material and the treatment conditions.

iii. Regeneration of plants is not very easy, particularly when protoplasts are used.

2. Particle Bombardment (Biolistics):

Particle (or micro projectile) bombardment is the most effective method for gene transfer, and creation of transgenic plants. This method is versatile due to the fact that it can be successfully used for the DNA transfer in mammalian cells and microorganisms.

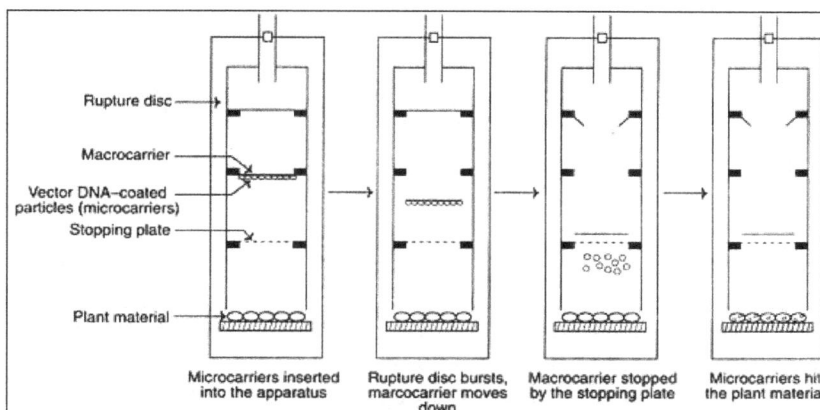

A diagrammatic representation of Particle Bombardment (biolistics)System for Gene Transfer in Plants.

The micro projectile bombardment method was initially named as biolistics by its inventor Sanford. Biolistics is a combination of biological and ballistics. There are other names for this technique- particle gun, gene gun, bio blaster. A diagrammatic representation of micro projectile bombardment system for the transfer of genes in plants is depicted, and briefly described below.

Micro carriers (micro projectiles), the tungsten or gold particles coated with DNA, are carried by macro carriers (macro projectiles). These macro-carriers are inserted into the apparatus and pushed downward by rupturing the disc.

The stopping plate does not permit the movement of macro carrier while the micro carriers (with DNA) are propelled at a high speed into the plant material. Here the DNA segments are released which enter the plant cells and integrate with the genome.

Plant Material used in Bombardment

Two types of plant tissue are commonly used for particle bombardment:

1. Primary explants which can be subjected to bombardment that are subsequently induced to become embryo genic and regenerate.

2. Proliferating embryonic tissues that can be bombarded in cultures and then allowed to proliferate and regenerate.

In order to protect plant tissues from being damaged by bombardment, cultures are maintained on high osmoticum media or subjected to limited plasmolysis.

Transgene Integration in Bombardment

It is believed (based on the gene transfer in rice by biolistics) that the gene transfer in particle bombardment is a two stage process.

1. In the pre-integration phase, the vector DNA molecules are spliced together. This results in fragments carrying multiple gene copies.

2. Integrative phase is characterized by the insertion of gene copies into the host plant genome.

The integrative phase facilitates further transgene integration which may occur at the same point or a point close to it. The net result is that particle bombardment is frequently associated with high copy number at a single locus. This type of single locus may be beneficial for regeneration of plants.

Success of Bombardment

The particle bombardment technique was first introduced in 1987. It has been successfully used for the transformation of many cereals, e.g. rice, wheat, maize. In fact, the first commercial genetically modified (CM) crops such as maize containing Bt-toxin gene were developed by this approach.

A selected list of the transgenic plants (developed by biolistics) along with the sources of the plant materials used is given in table.

Table: Selected list of transgenic plants (along with cell sources) developed by microprojectlle bombardment

Plant	Cell source(s)
Rice	Embryonic callus, immature Zygotic embryos
Wheat	Immature zygotic embryos
Sorghum	Immature zygotic embryos
Corn	Embryonic cell suspension, immature zygotic embryos
Barley	Cell suspension, immature zygotic embryos
Banana	Embryonic cell suspension
Sweet potato	Callus cells
Cotton	Zygotic embryos
Grape	Embryonic callus
Peas	Zygotic embryos
Peanut	Embryonic callus
Tobacco	Pollen
Alfalfa	Embryonic callus

Factors Affecting Bombardment

Several attempts are made to study the various factors, and optimize the system of particle bombardment for its most efficient use. Some of the important parameters are described.

Nature of Micro Particles

Inert metals such as tungsten, gold and platinum are used as micro particles to carry DNA. These particles with relatively higher mass will have a better chance to move fast when bombarded and penetrate the tissues.

Nature of Tissues/Cells

The target cells that are capable of undergoing division are suitable for transformation. Some more details on the choice of plant material used in bombardment are already given.

Amount of DNA

The transformation may be low when too little DNA is used. On the other hand, too much DNA may result is high copy number and rearrangement of transgenes. Therefore, the quantity of DNA used should be balanced. Recently, some workers have started using the chemical aminosiloxane to coat the micro particles with low quantities of DNA adequate enough to achieve high efficiency of transformation.

Environmental Parameters

Many environmental variables are known to influence particle bombardment. These factors (temperature, humidity, photoperiod etc.) influence the physiology of the plant material, and consequently the gene transfer. It is also observed that some explants, after bombardment may require special regimes of light, humidity, temperature etc.

The technology of particle bombardment has been improved in recent years, particularly with regard to the use of equipment. A commercially produced particle bombardment apparatus namely PDS-1000/HC is widely used these days.

Advantages of particle bombardment:

    i.   Gene transfer can be efficiently done in organized tissues.

    ii.   Different species of plants can be used to develop transgenic plants.

Limitations of particle bombardment:

    i.   The major complication is the production of high transgene copy number. This may result in instability of transgene expression due to gene silencing.

    ii.   The target tissue may often get damaged due to lack of control of bombardment velocity.

    iii.   Sometimes, undesirable chimeric plants may be regenerated.

3. Microinjection:

Microinjection is a direct physical method involving the mechanical insertion of the desirable DNA into a target cell. The target cell may be the one identified from intact cells, protoplasts, callus, embryos, meristems etc. Microinjection is used for the transfer of cellular organelles and for the manipulation of chromosomes.

The technique of microinjection involves the transfer of the gene through a micropipette (0.5-10.0 pm tip) into the cytoplasm/nucleus of a plant cell or protoplast. While the gene transfer is done, the recipient cells are kept immobilized in agarose embedding, and held by a suction holding pipette.

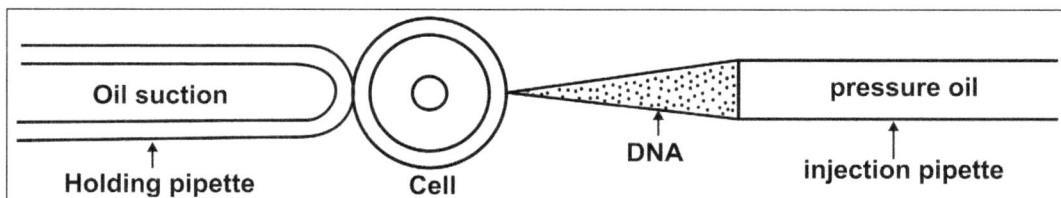

Microinjection of DNA by holding pipette method.

As the process of microinjection is complete, the transformed cell is cultured and grown to develop into a transgenic plant. In fact, transgenic tobacco and Brassica napus have been developed by this approach. The major limitations of microinjection are that it is slow, expensive, and has to be performed by trained and skilled personnel.

4. Liposome-Mediated Transformation:

Liposomes are artificially created lipid vesicles containing a phospholipid membrane. They are successfully used in mammalian cells for the delivery of proteins, drugs etc. Liposomes carrying genes can be employed to fuse with protoplasts and transfer the genes.

The efficiency of transformation increases when the process is carried out in conjunction with polyethylene glycol (PEG). Liposome-mediated transformation involves adhesion of liposomes to the protoplast surface, its fusion at the site of attachment and release of plasmids inside the cell.

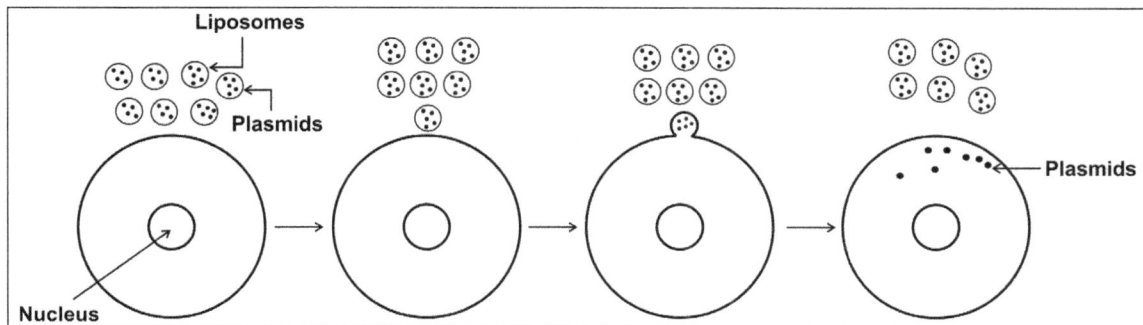

Adiagrammatic representation of Fusion of Plasmid-Filled Liposomes with Protoplasts.

Advantages of liposome fusion:

i.   Being present in an encapsulated form of liposomes, DNA is protected from environmental insults and damage.

ii. DNA is stable and can be stored for some time in liposomes prior to transfer.

iii. Applicable to a wide range of plant cells.

iv. There is good reproducibility in the technique.

Limitations of Liposome Fusion:

The major problem with liposome-mediated transformation is the difficulty associated with the regeneration of plants from transformed protoplasts.

5. Silicon Carbide Fibre-Mediated Transformation:

The silicon carbide fibres (SCF) are about 0.3-0.6 pm in diameter and 10-100 pm in length. These fibres are capable of penetrating the cell wall and plasma membrane, and thus can deliver DNA into the cells. The DNA coated silicon carbide fibres are vortexed with 'plant material (suspension culture, calluses). During the mixing, DNA adhering to the fibres enters the cells and gets stably integrated with the host genome. The silicon carbide fibres with the trade name Whiskers are available in the market.

Advantages of SCF-Mediated Transformation:

i.   Direct delivery of DNA into intact walled cells. This avoids the protoplast isolation.

ii. Procedure is simple and does not involve costly equipment.

Disadvantages of SCF-mediated transformation:

    i.   Silicon carbide fibres are carcinogenic and therefore have to be carefully handled.

    ii. The embryonic plant cells are hard and compact and are resistant to SCF penetration.

In recent years, some improvements have been made in SCF-mediated transformation. This has helped in the transformation of rice, wheat, maize and barley by using this technique.

Chemical Gene Transfer Methods

1. Polyethylene glycol (PEG)-mediated transfer:

Polyethylene glycol (PEG), in the presence of divalent cations (using Ca2+), destabilizes the plasma membrane of protoplasts and renders it permeable to naked DNA. In this way, the DNA enters nucleus of the protoplasts and gets integrated with the genome.

The procedure involves the isolation of protoplasts and their suspension, addition of plasmid DNA, followed by a slow addition of 40% PEG-4000 (w/v) dissolved in mannitol and calcium nitrate solution. As this mixture is incubated, protoplasts get transformed.

Advantages of PEG-mediated transformation:

    i.   A large number of protoplasts can be simultaneously transformed.

    ii. This technique can be successfully used for a wide range of plant species.

Limitations of PEG-mediated transformation:

    i.   The DNA is susceptible for degradation and rearrangement.

    ii. Random integration of foreign DNA into genome may result in undesirable traits.

    iii. Regeneration of plants from transformed protoplasts is a difficult task.

2. Deae Dextran-Mediated transfer:

The desirable DNA can be complexed with a high molecular weight polymer diethyl amino ethyl (DEAE) dextran and transferred. The major limitation of this approach is that it does not yield stable trans-formants.

Calcium Phosphate Co-precipitation-mediated Transfer

The DNA is allowed to mix with calcium chloride solution and isotonic phosphate buffer to form DNA-calcium phosphate precipitate. When the actively dividing cells in culture are exposed to this precipitate for several hours, the cells get transformed. The success of this method is dependent on the high concentration of DNA and the protection of the complex precipitate. Addition of dimethyl sulfoxide (DMSO) increases the efficiency of transformation.

DNA Imbibition by Cells/Tissues

Some workers have seriously tried to transform cells by incubating cell suspensions, tissues, embryos and even seeds with DNA. The belief is that the DNA gets imbibed, and the cells get transformed. DNA imbibition approach has met with little or no success.

Transgenic Plants

Transgenic plants or genetically modified plants are plants whose DNA is modified using genetic engineering techniques. In most cases the aim is to introduce a new trait to the plant which does not occur naturally in this species. Examples include resistance to certain pests, diseases or environmental conditions, or the production of a certain nutrient or pharmaceutical agent.

A transgenic crop plant contains a gene or genes which have been artificially inserted instead of the plant acquiring them through pollination. The inserted gene sequence, known as the transgene, may come from another unrelated plant or from a completely different species. Plants containing transgenes are often called genetically modified or GM crops.

Reasons for Making Transgenic Crop Plants

i. The process of transgenic plant development primarily aims at assembling a combination of genes in a crop plant which will make it as useful and productive as possible. Depending on where and for what purpose the plant is grown, desirable genes may provide features such as higher yield or improved quality, pest or disease resistance, or tolerance to heat, cold and drought.

ii. Combining the best genes in one plant is a long and difficult process, especially as traditional plant breeding has been limited to artificially crossing plants within the same species or with closely related species to bring different genes together.

   For example, a gene for protein in soybean could not be transferred to a completely different crop such as corn using traditional techniques. Transgenic technology enables plant breeders to bring together in one plant useful genes from a wide range of living sources, not just from within the crop species or from closely related plants.

iii. This technology provides the means for identifying and isolating genes controlling specific characteristics in one kind of organism, and for moving copies of those genes into another quite different organism, which will then also have those characteristics.

iv. This powerful tool enables plant breeders to do what they have always done-to generate more useful and productive crop varieties containing new combinations of genes-but it expands the possibilities beyond the limitations imposed by traditional cross-pollination and selection techniques.

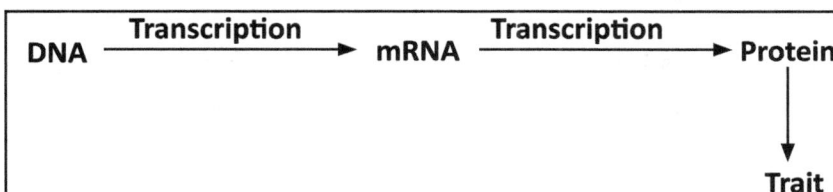

The Universal Concept of Expression of Biological Traits.

Fundamentals of Transgenic Plant Development

The underlying reason that transgenic plants can be constructed is the universal presence of DNA (deoxyribonucleic acid) in the cells of all living organisms. This molecule stores the organism's genetic information and orchestrates the metabolic processes of life. Genes are discrete segments of DNA that encode the information necessary for assembly of a specific protein.

A specific protein (or an enzyme) encodes for a particular trait. In the production of a transgenic plant our primary aim is to transfer a foreign gene, encoding for some novel traits, into the genome of the plant stably.

After transferring we also need the transgene to integrate and express in the plant's cells. This process as a whole generates a new variety of plant which is new in its own kind and interests us in its massive large scale cultivation.

Steps Involved in the Production of Transgenic Plants

The fundamental steps involved in the transgenic plant production are as follows:

Step 1: Identifying, Isolation and Cloning of Genes for Agriculturally Important Traits:

The very first step in the generation of transgenic plant is to identify and isolate the novel transgene that we want to transfer into the genome of the target plant. Usually, identifying a single gene involved with a trait is not sufficient. We also have to understand how the gene is regulated, what other effects it might have on the plant, and how it interacts with other genes active in the same biochemical pathway.

Step 2: Designing Gene Construct for Insertion:

After entering the plant cell the transgene must inter-grate into the genome of the plant stably and express itself successfully so as to produce higher amount of transgenic protein which will be indirectly reflected in the trait controlled by it.

To achieve this we have to design a "gene construct" or gene-set, having all the DNA segments necessary to achieve the integration and expression of the transgene. Once a gene has been isolated and cloned (amplified in a bacterial vector), it must undergo several modifications before it can be effectively inserted into a plant.

A gene-set which will be transferred to the target plant has following segments:

1. A Promoter Sequence: This must be added for the gene to be correctly expressed (i.e., translated into a protein product). The promoter is the on/off switch that controls when and where in the plant the gene will be expressed. To date, most promoters in transgenic crop varieties have been "constitutive", i.e., causing gene expression throughout the life cycle of the plant in most tissues.

The most commonly used constitutive promoter is CaMV 35S, from the cauliflower mosaic virus, which generally results in a high degree of expression in plants. Other promoters are more specific and respond to cues in the plant's internal or external environment. An example of a light-inducible promoteris the promoter from the cab gene, encoding the major chlorophyll a/b binding protein.

2. The Transgene: Sometimes, the transgene is modified to achieve greater expression in a plant.

For example, the Bt gene for insect resistance is of bacterial origin and has a higher percentage of A-T nucleotide pairs compared to plants, which prefer G-C nucleotide pairs. In a clever modification, researchers substituted A-T nucleotides with G-C nucleotides in the Bt gene without significantly changing the amino acid sequence. The result was enhanced production of the gene product in plant cells.

3. Termination Sequence: This signals to the cellular machinery that the end of the gene sequence has been reached.

4. A Selectable Marker Gene: This is added in order to identify plant cells or tissues that have successfully integrated the transgene. This is necessary because achieving incorporation and expression of transgenes in plant cells is a rare event, occurring in just a few per cent of the targeted tissues or cells.

Selectable marker genes encode proteins that provide resistance to agents that are normally toxic to plants, such as antibiotics or herbicides. Only plant cells that have integrated the selectable marker gene will survive when grown on a medium containing the appropriate antibiotic or herbicide.

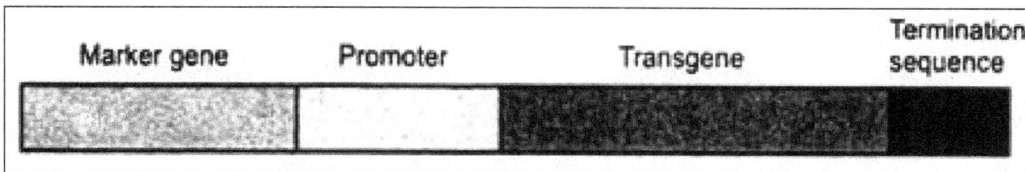

The Gene Construct that has to be transgerred into the target plant.

The essential features of an ideal reporter gene are:

1. Lack of endogenous activity in plant cells of the concerned enzyme,

2. An efficient and easy detection, and

3. A relatively rapid degradation of the enzyme.

The commonly used selectable marker genes include those conferring resistance to the antibiotics kanamycin (nptII, encoding neomycin phosphotransferase) and hygromycin (hptIV, encoding hygromycin phosphotransferase, isolated from E. coli); and broad range herbicides glyphosate (modified versions of the enzyme EPSPS, 5-enolpyruvate shikimate-3-phosphate synthase, isolated from E. coli or Salmonella typhimurium), phosphinothricin (bar, isolated from Streptomyces hygroscopicus, codes for phophinothricin acetyltransferase), etc.

Step 3: Transforming Target Plants with the Gene Construct:

There are two ways of genetically transforming the target plant:

• Vector mediated gene transfer, and

• Vector less or direct gene transfer.

Step 4: Selection of The Transgenic Plant Tissue/Cells:

Following the gene insertion process, plant tissues are transferred to a selective medium containing an antibiotic or herbicide, depending on which selectable marker was used. Only plants expressing the selectable marker gene will survive and it is assumed that these plants will also possess the transgene of interest.

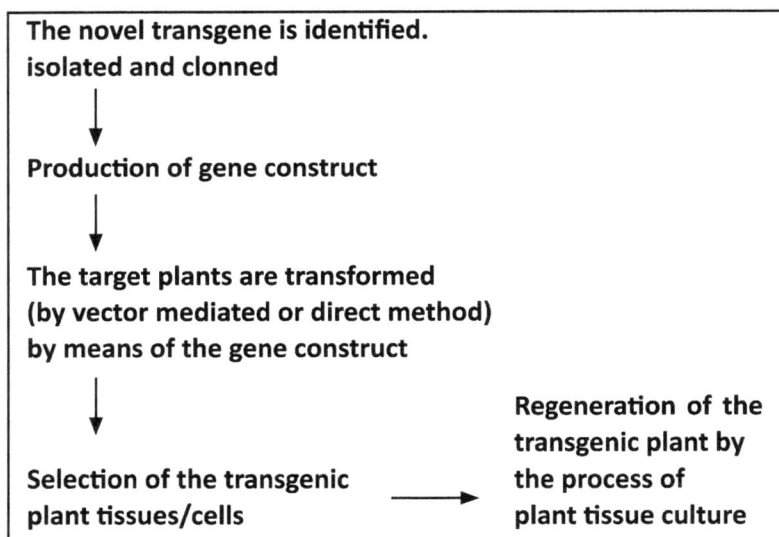

The novel transgene is identified.
isolated and clonned

↓

Production of gene construct

↓

The target plants are transformed
(by vector mediated or direct method)
by means of the gene construct

↓

Selection of the transgenic   →   Regeneration of the
plant tissues/cells   transgenic plant by
                                            the process of
                                            plant tissue culture

Basics involved in the Production of Transgenic Plants

Step 5: Regeneration of the Transgenic Plants:

To obtain whole plants from transgenic tissues, they are grown under controlled environmental conditions in a series of media containing nutrients and hormones by the process of plant tissue culture.

Table: Promoters used in construction of gene construct.

Promoter	Source	Relative activity
35S	CaMV 35S RNA gene	Constitutive, high activity; most commonly used commonly used in dicots
35S + Adh 1-I 1	35S promoter + first intron of maize Adh 1 gene	Enhanced promoter activity; constitutive
35S + sh 1-I 1	35S promoter + first intron of maize shrunken 1 gene	Better than 35S + Adh1-I 1 in mono cots; constitutive
Adh 1	Promoter of alcohol dehydrogenase gene of maize	Moderate activity in cereals; anaerobic expression
Emu	Modified from Adh 1 promoter and its first intron	Moderate activity in cereals; anaerobic expression
Act1 + Act-I 1	Rice actin gene + its first intron	Moderate activity; constitutive
Ubi 1 + Ubi 1-I 1	Maize ubiquitin 1 gene promoter + its first (or sixth) intron	High activity in cereals; constitutive
Vicilin promoter	Pea vicilin storage protein gene	Seed-specific promoter

Step 6: Integration of the Transgene in the Genome of the Target Plant:

In general, transgenes integrate at random sites in any of the chromosomes of the genome of host cells. Usually, in a given cell, integration occurs at a single location. As a result, different cells may be expected to show integration of the transgene at different chromosomal locations.

The number of copies integrated per genome ranges from one to several hundred. In general, multiple copies are integrated when large amounts of DNA are used for transfection, while single copies are integrated with smaller amounts.

When multiple copies are integrated, they are mostly integrated at one site joined to each other head-to-tail, i.e., as a concatemer. However, in a small proportion of cases, the multiple copies are located at several sites in the same genome.

The mechanism of random integration is not known. The entire gene construct, including the vector DNA, becomes integrated. When two different gene constructs are mixed and used for transfection, they tend to be integrated together at the same site; this is known as co-transfection. The sequences flanking a gene on either side influence the expression of this gene.

Therefore, the same transgene integrated at different locations in the genome may show different levels of expression; this is known as position effect. Transgene integration frequently leads to various forms of rearrangements, e.g., duplication, deletion, etc., near the site of integration.

If these changes are large enough, the host gene located at the site of integration may become non-functional. A host gene would also become non-functional if the transgene becomes integrated within the coding region of this gene. When integration of a transgene leads to the loss of function of a host gene, it is called insertional mutagenesis; it often produces aberrant phenotypes.

Analysis of Transgene Integration

The integration of transgene into the genome is confirmed by Southern hybridization of genomic DNA extracted from the considered transgenic individuals. The DNA is digested with a suitable restriction enzyme prior to electrophoresis.

By choosing appropriate restriction enzymes for DNA digestion, not only the integration of transgene can be established beyond doubt, but information on the number of copies per cell, the orientations of tandemly arranged copies and the presence of single or multiple integration sites is also obtained from Southern hybridization. All the individuals that give positive result with Southern hybridization are regarded as confirmed transgenic.

Detection of mRNA Expression

The mRNAs produced by transgenes is most readily detected if they are with unique sequences, which have no counterparts among those produced by the host genome. A high purity RNA preparation is obtained from the appropriate tissue of transgenic individuals, and is subjected to RNA dot blot hybridization with a radioactive probe specific for the transgene.

Alternatively, the RNA preparation may be used for northern hybridization, which provides additional information on transcript size as well.

Inheritance of Transgenes

The transgenes which are stably integrated are inherited in a Mendelian fashion. They are usually dominant. Instability may occur due to point mutation, like methylation, or rearrangements of the T-DNA region. In addition, homologous recombination between copies of the transgene inserted in the same nucleus can also lead to instability of the gene.

Applications of Transgenic Plants

Resistance to Biotic Stresses

Genetic engineering of plants has led to the development of crops with increased resistance to biotic stresses which is described in three major categories:

1. Insect (PEST) Resistance:

It is estimated that about 15% of the world's crop yield is lost to insects or pests. A selected list of the common insects and the crops damaged is given in table.

A selected list of common insect pests along with the major crops damaged by them		
Common name of pest	Botanical name	Crop(s) damaged
Cotton bollworm	Helicoverpa zea	Cotton
Cotton leafworm	Spodoptera littiralis	Rice, maize, cotton, tobacco
Tobacco hornworm	Manduca sexta	Tabacco, tomato, potato
European corn borer	Ostrinia nubilalis	Maize
Locust	Locusta migratoria	Grasses
Tobacco budworm	Heliothis virescens	Tobacco, cotton
Tomato fruitworm	Heliothis armigera	Tomato, cotton
Cowpea seed beetle	Callosobruchus maculatus	Cowpea, soybean
Colorado beetle	Leptinotarsa decemlineata	Potato
Brown plant hopper	Nilaparvata lugens	Rice

The damage to crops is mainly caused by insect larvae and to some extent adult insects.

The majority of the insects that damage crops belong to the following orders (with examples):

  i.   Lepidoptera (bollworms)

  ii. Coleoptera (beetles)

  iii. Orthoptera (grasshoppers)

  iv. Homoptera (aphids)

Till some time ago, chemical pesticides are the only means of pest control.

Scientists have been looking for alternate methods of pest control for the following reasons (i.e. limitations of pesticide use):

i.   About 95% of the pesticide sprayed is washed away from the plant surface and accumulates in the soil.

ii.   It is difficult to deliver pesticides to vulnerable parts of plants such as roots, stems and fruits.

iii. Chemical pesticides are not efficiently degraded in the soil, causing environmental pollution.

iv.   Pesticides, in general, are toxic to non- target organisms, particularly humans and animals.

It is fortunate that scientists have been able to discover new biotechnological alternatives to chemical pesticides thereby providing insect resistance to crop plants. Transgenic plants with insect resistance transgenes have been developed. About 40 genes obtained from microorganisms of higher plants and animals have been used to provide insect resistance in crop plants. Some of the approaches for the bio-control of insects.

Resistance Genes from Microorganisms

Bacillus Thuringiensis (Bt) Toxin

Bacillus thuringiensis was first discovered by Ishiwaki in 1901, although its commercial importance was ignored until 1951. B. thuringiensis is a Gram negative, soil bacterium. This bacterium produces a parasporal crystalline proteinous toxin with insecticidal activity. The protein produced by B .thuringiensis is referred to as insecticidal crystalline protein (ICP). ICPs are among the endotoxins produced by sporulating bacteria, and were originally classified as δ-endotoxins (to distinguish them from other classes of α-, β- and γ-endotoxins).

Bt Toxin Genes

Several strains of B. thuringiensis producing a wide range of crystal (cry) proteins have been identified. Further, the structure of cry genes and their corresponding toxin (δ-endotoxin) products have been characterized. The cry genes are classified (latest in 1998) into a large number of distinct families (about 40) designated as cry 1 - cry 40, based on their size and sequence similarities. And within each family, there may be sub-families. Thus, the total number of genes producing Bt toxins (Cry proteins) is more than 100.

There are differences in the structure of different Cry proteins, besides certain sequence similarities. The molecular weights of Cry proteins may be either large (~130 KDa) or small (~70KDa). Despite the differences in the Cry proteins, they share a common active core of three domains.

Mode of Action of Cry Proteins

Most of the Bt toxins (Cry proteins) are active against Lepidopteran larvae, while some of them are specific against Dipteran and Coleopteran insects. The pro-toxin of Cry I toxin group has a molecular mass of 130 kilo Daltons (130 KDa).

When this parasporal crystal is ingested by the target insect, the pro-toxin gets activated within its gut by a combination of alkaline pH (7.5 to 8.5) and proteolytic enzymes. This results in the conversion of pro-toxin into an active toxin with a molecular weight of 68 KDa.

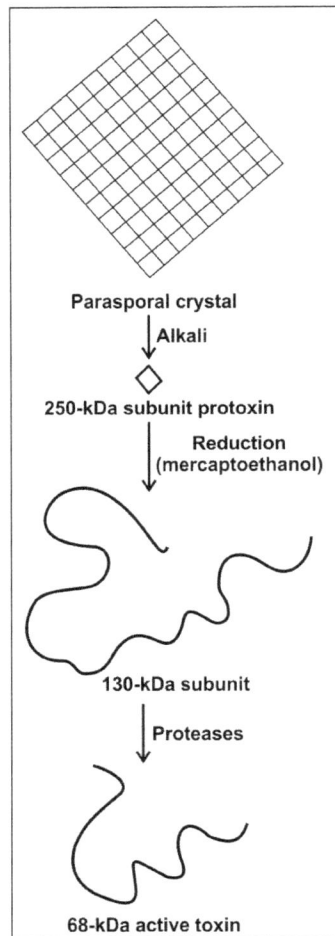

A diagrammatic representation of the Formation of Active Toxin from the
parasporal crystal of Bacillus thuringiensis.

The active form of toxin protein gets itself inserted into the membrane of the gut epithelial cells of the insect. This result in the formation of ion channels through which there occurs an excessive loss of cellular ATP. As a consequence, cellular metabolism ceases, insect stops feeding, and becomes dehydrated and finally dies.

Some workers in the recent years suggest that the Bt toxin opens cation-selective pores in the membranes, leading to the inflow of cations into the cells that causes osmotic lysis and destruction of epithelial cells (and finally the death of insect larvae). The Bt toxin is not toxic to humans and animals since the conversion of pro-toxin to toxin requires alkaline pH and specific proteases (These are absent humans and animals).

Bt Toxin as Bio-pesticide

Preparations of Bt spores or isolated crystals have been used as organic bio-pesticide for about 50 years.

This approach has not met with much success for the following reasons:

    i.   Low persistence and stability (sunlight degrades toxin) of the toxin on the surface of plants.

ii. The Bt toxin cannot effectively penetrate into various parts of plants, particularly roots.

iii. Cost of production is high.

Bt-based Genetic Transformation of Plants

It has been possible to genetically modify (CM) plants by inserting Bt genes and provide pest resistance to these transformed plants. For an effective pest resistance, the bacterial gene in transgenic plants must possess high level expression. This obviously means that the transgene transcription should be under the effective control of promoter and terminator sequences. The early attempts to express cry 1A and cry 3A proteins under the control of CaMV 35S or Agrobacterium T-DNA promoters resulted in a very low expression in tobacco, tomato and potato plants.

Modification of Bt cry 1A Gene

The wild type transgene Bt cry 1A(b) was found to express at a very low levels in transgenic plants. The nucleotide sequence of this gene was modified (G + C content altered, several polyadenylation signals removed, ATTTA sequence deleted etc). With appropriate sequence changes, an enormous increase (about 100 fold) in the Bt toxin product formation was observed.

The transgenic Bt crops that were found to provide effective protection against insect damage were given approval for commercial planting by USA in the mid-1990s. Some biotechnological companies with their own trade names introduced several transgenic crops into the fields. Among these, only maize and cotton Bt crops are currently in use in USA. The other genetically modified plants met with failure for various reasons.

Bt-based Genetically Modified Crop Plants plants developed for commercial use			
Crop	Trade name	Bt protein	Resistane to insect(s)
Cotton	Bollgard	Cry 1Ac	Cotton bollworm, tobacco budworm
Maize	Yield gard knockout	Cry 1Ab	European conr borer
Maize	Starlink	Cry 9c	European conr borer
Maize	Herculex I	Cry 1F	European conr borer
Maize	Bt – Xtra	Cry 1Ac	European conr borer
Potato	New-leaf	Cry 3A	Colarado beetle

Advantages of Transgenic Plants with Bt Genes:

i. Bt genes could be expressed in all parts of the plants, including the roots and internal regions of stems and fruits. This is not possible by any chemical pesticide.

ii. Toxic proteins are produced within the plants; hence they are environmental-friendly.

iii. Bt toxins are rapidly degraded in the environment.

Problem of Insect Resistance to Bt Crops

The major limitation of Bf-gene possessing transgenic plants is the development of Bt-resistant insects. The Bf toxin is a protein, and the membrane receptor (of the gut) through which the toxin mediates its action is also a protein. It is possible that the appropriate mutations in the insect gene coding for receptor protein may reduce the toxin binding and render it ineffective. This may happen within a few generations by repeated growing of Bt crops.

Several approaches are made to avoid the development of resistance in insects:

    i.   Introduction of two different Bf toxin genes for the same target insect.

    ii.   Development of transgenic plants with two types of insect resistance genes e.g. Bt gene and proteinase inhibitor gene.

    iii.   Rotating Bt crops with non-Bf crops may also prevent the build-up of resistance in insect population.

Environmental Impact of Bt Crops

The most serious impact of Bt crops on environment is the build-up of resistance in the pest population. In 1999, another issue was brought to light about Bt crops. It was reported that the pollen from Bt maize might be toxic to the larvae of Monarch butterfly. This generated considerable opposition to Bt crops by the public, since Monarch butterfly is one of the most colorful natives in USA.

It was later proved that the fears about the impact of Bt crops on the monarch butterfly were without the required scientific evidence. The lesson learnt from the monarch butterfly episode is that the risks of CM crops should be thoroughly assessed before they are reported.

Usage of Bt

The usage Bt is commonly used for a transgenic crop with a cry gene e.g. Bt cotton. In the same way, Cry proteins are also referred to as Bt proteins. It may also be stated here that the authors use four different names for the same group of proteins-δ-endotoxin, insecticidal crystal protein (ICP), Cry and now Bt.

Resistance Genes from Other Microorganisms

There are certain other insect resistant genes from other microorganisms. Some of the important ones are listed:

    1.   Cholesterol oxidase of Strepotmyces culture filtrate was found to be toxic to boll weevil larvae. Cholesterol oxidase gene has been introduced into tobacco to develop a transgenic plant.

    2.   Isopentenyl transferase gene from Agrobacterium tumefaciens has been introduced into tobacco and tomato. This gene codes for an important enzyme in the synthesis of cytokinin. The transgenic plants with this transgene were found to reduce the leaf consumption by tobacco hornworm and decrease the survival of peach potato aphid.

Resistance Genes from Higher Plants

Certain genes from higher plants were also found to result in the synthesis of products possessing insecticidal activity. Some authors regard them as non-Bt insecticidal proteins. A selected list of plant insecticidal (non-Bt) genes used for developing transgenic plants with insect resistance is given.

A selected list of plant insecticidal (non-Bt) genes used for developing transgenic plants with insect resistance.

Plant gene	Transgenic plants(s)	Encoded protein	Resistance to insect(s)
Protease inhibitors			
CpTi	Potato, apple, rice, Sunflower, wheat, tomato	Trypsin	Coleoptera, lepidoptera
Cll	Tobacco, potato	Serine protease	Coleoptera, lepidoptera
Pl-lV	Potato, tobacco	Serine protease	Lepidoptera
OC-1	Tobacco, oilseed rape	Cysteine peotease	Coleoptera, Homoptera
CMe	Tobacco	Trypsin	Lepidoptera
α-Amylase inhibitors			
α-A1-Pv	Pea, tobacco	α-Amylase	Coleoptera
WMAI-1	Tobacco	α-Amylase	Lepidoptera
Lectins			
GNA	Potato, rice, sugarcane, sweet potato, tobacco	Lectin	Homoptera, Lepidoptera
WGA	Maize	Agglutin	Lepidoptera, Coleoptera
Others			
BCH	Potato	Chitinase	Homoptera, Lepidoptera
TOC	Tobacco	Tryptophan decarboxylase	Homoptera

Proteinase (Protease) Inhibitors

Proteinase inhibitors are the proteins that inhibit the activity of proteinase enzymes. Certain plants naturally produce proteinase inhibitors to provide defence against herbivorous insects. This is possible since the inhibitors when ingested by insects interfere with the digestive enzymes of the insect. This results in the nutrient deprivation causing death of the insects. It is possible to control insects by introducing proteinase inhibitor genes into crop plants that normally do not produce these proteins.

Cowpea Trypsin Inhibitor Gene

It was observed that the wild species of cowpea plants growing in Africa were resistant to attack by a wide range of insects. Research findings revealed that insecticidal protein was a trypsin inhibitor that was capable of destroying insects belonging to the orders Lepidoptera (e.g. Heliothis virescans), Orthaptera (e.g. Locusta migratoria) and Coleoptera (e.g. Anthonous grandis).

Cowpea trypsin inhibitor (CpTi) has no affect on mammalian trypsins; hence it is non-toxic to mammals. CpTi gene was introduced into tobacco, potato and oilseed rape for developing transgenic plants. Survival of insects and damage to plants were much lower in plants possessing CpTi gene.

Advantages of Proteinase Inhibitors:

 i. Many insects, not controlled by Bt, can be effectively controlled.

 ii. Use of proteinase gene along with Bt gene will help to overcome Bt resistance development in plants.

Limitations of proteinase inhibitors:

 i. Unlike Bt toxin, high levels of proteinase inhibitors are required to kill insects.

 ii. It is necessary that the expression of proteinase inhibitors should be very low in the plant parts consumed by humans, while the expression should be high in the parts of plants utilized by insects.

α-Amylase Inhibitors

The insect larvae secrete a gut enzyme a- amylase to digest starch. By blocking the activity of this enzyme by α-amylase inhibitor, the larvae can be starved and killed. α-Amylase inhibitor gene (α-AI-Pv) isolated from bean has been successfully transferred and expressed in tobacco. It provides resistance against Coleoptera (e.g. Zabrotes sub-fasciatus).

Lectins

Lectins are plant glycoproteins and they provide resistance to insects by acting as toxins. The lectin gene (CNA) from snowdrop (Calanthus nivalis) has been transferred and expressed in potato and tomato. The major limitations of lectin are that it acts only against piercing and sucking insect, and high doses are required.

Resistance Genes from Animals

Proteinase inhibitor genes from mammals have also been transferred and expressed in plants to provide resistance against insects, although the success in this direction is very limited. Bovine pancreatic trypsin inhibitor (BPTI) and α_1– antitrypsin genes appear to be promising to offer insect resistance to transgenic plants.

Insect Resistance through Copy Nature Strategy

Some of the limitations experienced by transferring the insecticidal genes (particularly Bt) and developing transgenic plants have prompted scientists to look for better alternatives. The copy nature strategy was introduced in 1993 (by Boulter) with the objective of insect pest control which is relatively sustainable and environmentally friendly.

The copy nature strategy for the development of insect -resistant transgenic plants has the following stages:

 i. Identification of leads: The first step in copy nature strategy is to identify the plants (from world over) that are naturally resistant to insect damage.

 ii. Isolation and purification of protein: The protein with insecticidal properties (from the resistant plants) is isolated and purified. The sequence of the protein is determined, and the gene responsible for its production identified.

3. Bioassay of isolated protein: The activity of the protein against the target insects is determined by performing a bioassay in the laboratory.

4. Testing for toxicity in mammals: It is absolutely necessary to test the toxicity of the protein against mammals, particularly humans. If the protein is found to have any adverse effect, the copy nature strategy should be discontinued.

5. Gene transfer: By the conventional techniques of genetic engineering, the isolated gene corresponding to the protein toxin is introduced into the crop plants.

6. Selection of transgenic plants: After the gene transfer, the transgenic plants developed should be tested for the inheritance and appropriate expression of transgene. The efficiency of insecticidal protein to destroy insects is also evaluated.

7. Evaluation for biosafety: Field trails have to be conducted to evaluate the crop yield, damage to insects, influence on the environment with respect to the transgenic plant. The copy nature strategy, though time consuming and many a times unsuccessful, takes into account the complex interplay in biological communities (between plants animals, microorganisms) and physical environment.

2. Virus Resistance:

Virus infections of crops may result in retarded cell division (hypoplasia), excessive cell division (hyperplasia), and cell death (necrosis). The overall effects of virus infections are growth retardation, lowered product yield and sometimes complete crop failure. The chemical methods used to control various plant pathogens will be ineffective with respect to plant viruses since the viruses are intracellular obligate parasites.

There are however, certain safe agricultural practices to control/reduce viral infections to plants:

i. Use of seeds that are virus – free.

ii. Control insects that spread plant viruses.

iii. Control weeds that serve as alternate hosts for viruses.

iv. Use cultivars that possess virus resistance.

It is possible to immunize plants against viral damages by expressing viral proteins in the plant cells. With the advances made in genetic engineering, it has become a reality to develop transgenic plants with virus resistance.

This is mostly done by employing virus-encoded genes – virus coat proteins, movement proteins, transmission proteins, satellite RNA, antisense RNAs and ribozymes. In recent years, some attempts are also made to provide virus resistance to plants by using animal genes.

Virus Coat Proteins

The virus coat protein-mediated approach is the most successful one to provide virus resistance to plants. It was in 1986, transgenic tobacco plants expressing tobacco mosaic virus (TMV) coat protein gene were first developed. These plants exhibited high levels of resistance to TMV. Excited by this remarkable success, scientists have worked with many more viruses (around 30 or so) and developed crops with virus coat protein-mediated protection.

A selected list of the virus resistant transgenic plants with sources of virus coat protein genes is given. The transgenic plant providing coat protein-mediated resistance to virus are rice, potato, wheat, tobacco, peanut, sugar beet, alfalfa etc. The viruses that have been used include alfalfa mosaic virus (AIMV), cucumber mosaic virus (CMV), potato virus X (PVX), potato virus Y (PVY), citrus tristeza virus (CTV) and R rice stripe virus (RSV).

A selected list of the virus- resistant transgenic plants with sources of virus protein coat genes.

Plant	Source(s) of virus coat protein gene
Tobacco	TMV, CMV, AIMV
Rice	RTSV, RSV, RYMV
Wheat	SBWMV, BYDV
Potato	PVX, PVY, PLRV
Squash	CMV, ZYMV
Sugar beet	BNYVV
Peanut	TSWV
Papaya	PRSV
Citrus	CTV
Alfalfa	AIMV

Advantages of Virus Coat Proteins

The coat protein gene from one virus sometimes provides resistance (cross protection) to some other viruses, which may be unrelated e.g. TMV of tobacco plant provides resistance to potato virus X, alfalfa mosaic virus and cucumber mosaic virus.

Limitation of Virus Coat Proteins

The virus coat protein-mediated protection is successful for viruses with single-stranded RNA genomes. However, this approach is of not much use for viruses with genomes containing double-stranded RNA and single-stranded DNA.

Mechanism of Action Of Virus Coat Proteins

As the transgenic plant expresses the gene for coat protein of a given virus, the ability of the same virus to infect the plants again is drastically reduced. Despite a remarkable success in the virus coat protein-mediated protection, the molecular mechanism of the protection is not clearly known.

Movement Proteins

As the virus infects the plant cells, its rapid spread through intercellular junctions (plasmadesmata) of vascular tissue occurs through the participation of movement proteins produced by the viruses. Good examples of movement proteins are 30KDa protein of tobacco mosaic virus (TMV) and 32KDa protein of brome mosaic virus (BMV).

Transgenic tobacco plants that express a mutated 30KDa movement protein have been developed. The TMV infection to these plants is much less. It is believed that the mutated movement protein competes with the wild-type TMV-coded protein thereby reducing the spread of the virus (TMV).

In recent years, a recombinant movement protein having the components of golden mosaic virus and African cassava mosaic virus have been developed. This protein effectively interferes with the spread of both the viruses. The advantage with movement protein strategy is that it is applicable to single-stranded DNA viruses (Gemini viruses) also.

Transmission Proteins

There is a good coordination and interaction between plant viruses and insect vectors for the spread of viruses from one plant to another. Certain viral-encoded transmission proteins do this job effectively. It is possible to produce mutated transmission proteins and block the spread of viruses. Thus, the spread of insect-transmitted viruses can be prevented by engineering crops to express a defective virus-transmission protein.

Satellite RNAs

Plant viral satellite RNAs are small RNA molecules that multiply in the host cells with the help of specific helper viruses. These satellite RNAs are encapsulated together with the respective helper viruses. In general, the presence of satellite RNAs reduces the severity of the viral disease and the symptoms, and thus reduces the effect of the virus.

Transgenic plants containing satellite sequences have been developed to provide resistance to virus diseases. One example is given here. When cucumber mosaic virus (CMV) infects pepper plants, severe symptoms appear. These symptoms can be minimized with higher plant yield when CMV is co-inoculated with a satellite RNA.

Satellite RNA approach is not widely used due to several limitations:

   i.   Some of the satellite RNA may increase the severity of disease symptoms in some plants.

   ii. Satellite RNAs mutate very rapidly which may sometimes result in a highly virulent agent.

   iii. Re-combinations between satellite RNAs have been detected. This may lead to serious consequences.

Antisense RNAs

The antisense RNA approach is designed to specifically interfere with virus replication. By use of genetic engineering, a complementary DNA Strand of a gene (DNA sequence) can be inserted in reverse orientation ($3' \rightarrow 5'$ as opposed to $5' \rightarrow 3'$) and this is referred to as antisense gene.

The mRNA produced by antisense gene is complementary to the mRNA synthesized by normal gene. As a result, both these mRNAs hybridize and thus the normal translation of mRNA is blocked. The net effect of employing an antisense gene into a cell is that it blocks a specific gene expression.

It is possible to introduce viral antisense genes into plants and produce mRNAs complementary to viral sequences involved in virus replication. The antisense mRNAs can block the replication of viruses. Initially, antisense RNA approach was carried out in single-stranded RNA viruses. The success of this approach however, was limited probably due to the following reasons.

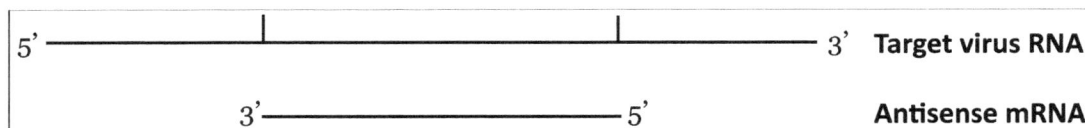

Hybridization of Antisense mRNA with virus RNA to block replication.

i.   High concentration of antisense mRNA may be required.

ii. Protein association with mRNA interferes with hybridization (between sense mRNA and antisense mRNA).

Antisense RNA approach may be more useful for DNA viruses. In fact, tomato golden mosaic virus (TGMV) replicase coding sequence was cloned in antisense orientation and introduced into tobacco plants. The transgenic tobacco plants expressed antisense RNA of TGMV replicase. These plants were resistant to TGMV infection.

Ribozymes

Ribozymes are small RNA molecules which promote the catalytic cleavage of RNA. For providing virus resistance, ribozymes in the form of antisense RNAs capable of cleaving the target viral (sense) RNAs have been developed. The ribozyme (antisense RNA) binds to a small sequence of viral RNA and splits.

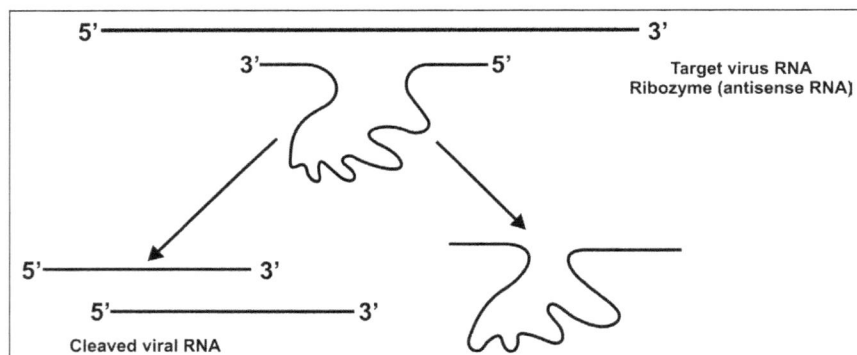

Action of Ribozymes as antieense RNAs to block virus replication.

In this way, it is possible to block the replication of viral RNA. However, the ribozymes approach has not been very successful in plants.

3. Fungal and Bacterial Diseases Resistance:

The plants do posses general defence systems against invading pathogens. This is however, not truly comparable with the immune system of the animals. Whenever there is a cellular damage caused by pathogens (fungi, bacteria) and plant pests, the general defence system of plants get geared up to provide some amount of protection to the plant. This natural disease resistance of plants is inadequate.

However, knowledge on the natural systems of plant resistance is useful for the biotechnological approaches to develop disease resistance.

Pathogenesis-Related (Pr) Proteins

To defend themselves against the invading pathogens (fungi and bacteria), plants accumulate low molecular weight proteins which are collectively regarded as pathogenesis-related (PR) proteins. The different types of PR proteins and their properties are given. Some of the most important ones are described.

Different types of pathogenesis related (PR) proteins produced in plants

Type	Properties
PR-1	Antifungal
PR-2	Endo β-1, 3-glucamases
PR-3	Emdochitinases
PR-4	Antifungal, endochitinase
PR-5	Antifungal, thaumatin-like proteins, osmotins
PR-6	Prosease inhibitors
PR-7	Endoprotease
PR-8	Chitinasefly sozyme
PR-9	peroxidases
PR-10	Rbonucleases
PR-11	Endochitinase acitivty
PR-12	Defensins
PR-13	Thionins
PR-14	Lipid transfer proteins (non-specific)

Chitinase

Chitin is a constituent of fungal cell walls which can be hydrolysed by the enzyme chitinase. Certain chitinase genes from plants have been isolated and characterized. A bacterial chitinase gene obtained from a soil bacterium (Serratia marcescens) was introduced and expressed in tobacco leaves. Some other workers isolated a chitinase gene from bean (Phaseolus vulgaris) and developed transgenic plants of tobacco and Brassica napus with this gene. The transformed tobacco plants were found to be resistant to infection of the pathogen Rhizoctonia solani. In case of B. napus, the protection however, was comparatively less.

Glucanase

Glucanase is another enzyme that degrades the cell wall of many fungi. The most widely used glucanase is β-1, 4-glucanase. The gene encoding for β-1, 4-glucanase has been isolated from barley, introduced, and expressed in transgenic tobacco plants. This gene provided good protection against soil-borne fungal pathogen Rhizoctonia solani.

The resistance to fungal pathogens is much higher if both chitinase and glucanase producing genes are present in transgenic plants. By this approach, fungal resistant tobacco, tomato and carrot have been developed.

Ribosome-In Activating Proteins (RIPs)

Ribosome-inactivating proteins offer protection against fungal infections. They act on the large rRNA of eukaryote and prokaryote ribosomes (remove an adenine residue from a specific site), and thus inhibit protein biosynthesis.

Certain RIPs that do not inhibit plant ribosomes were identified and the corresponding genes have been used to develop transgenic plants e.g. Type-I barley RIP is used to provide resistance to fungal infections. Some authors use the term antimicrobial proteins to RIPs. The other examples of antimicrobial proteins are lectins, defensins, lysozyme, thionins etc.

Lysozyme

Lysozyme degrades chitin and peptidoglycan of cell wall and in this way fungal infection can be reduced. Transgenic potato plants with lysozyme gene providing resistance to Eswinia carotovora have been developed.

Defensins

Defensins are antimicrobial peptides (26-50 amino acid residues) found in all the plant cells. They attack the microbial plasma membrane; however this is not adequate to provide resistance to pathogens. In recent years, an artificial defensin gene has been developed and introduced into potatoes. These potatoes developed resistance to the bacterium Eswinia carotovora.

Thionins

Thionin proteins also offer protection against bacteria. Thionin coding genes have been introduced into tobacco and the transgenic plant so developed showed resistance to Pseudomonas syringae.

Phytoalexins

Phytoalexins are secondary metabolites produced in plants in response to infection. They are low-molecular weight and antimicrobial in nature. The phytoalexins usually present in specialized cells or organelles are mobilized when infection occurs. Further, during infection there occurs induction of genes for increased production of phytoalexins.

Stilbene synthase is a key enzyme for the synthesis of a common phytoalexin. The gene coding stilbene synthase has been isolated from peanut and introduced into tobacco, rice and Brassica napus. The transgenic plants carrying stilbene synthase gene were resistant against some fungi. A selected list of transgenic plants developed, along with the genes transferred and the controlled pathogens is given in table.

A selected list of transgenic plants developed along with the genes transgerred and the resistance provided against the pathogens (fungus/bacterium).

Crop	Gene(s) transferred	Resistance against pathogen
Pathogenesis-related (PR) proteins		
Tobacco	Chitinase fro bacterium	Alternaria longipes
	(serratia marcescens)	
	bean chitinase	Rhizoctonia solani, Phytophthora parasitica
	Chitinase and 1, 3-β glucanase	Cercospora nicotinae
Rice	Chitinase	Rhizoctonia solani
Carrot	Chitinase and 1, 3-β glucanase	Alernaria dauci, A. radicina
Tomato	Chitinase and 1, 3-β glucanase	Fusarium oxysporum
Brassica napus	Chitinase	Rhizoctonia solani
Antimicrobial proteins		
Tobacco	Barly ribosome inactivating protein	Rhizoctonia solani
Tobacco	Defensin from radish	Alternaria longipes
Tobacco	α-Thionin gene from barley	Pseudomonas syringae
Potato	Bacteriophage T-4 lysozyme	Erwinia carotovora
phytoalexins		
Rice	Stilbene synthase	Pyricufaria oryzae
Tobacco	Stilbene synthase	Botrytis cinerea

Nematode Resistance

Nematodes are simple worms found in the soil. They possess a complete digestive tract. The annual crop loss of the world due to nematode (roundworm) infestation is very high.Some workers have identified and cloned a nematode resistance gene from wild beet plants.

It is proposed that this gene encodes a protein that detects the pests (nematodes) and triggers a defensive reactions in the plant. It is believed that some chemical compounds that destroy the gut of the nematode are produced. Attempts were also made to transfer the nematode resistance gene to sugar beet. Not much success was reported, the major limitation being the difficulty in cultivating the gene-altered cells of sugar beet.

Resistance to Abiotic Stresses

Plants are constantly being subjected to environmental stresses that may result in deterioration of crop plants, and a very low or even no yield. Plants are dependent on the subtle internal mechanisms for tolerance of various stresses.

The in situ tolerance of crop plants, whenever present, is inadequate and therefore cannot give protection against the stresses. A wide range of strategies are required to engineer plants against a particular type of stress tolerance.

Herbicide Resistance

Weeds (wild herbs) are unwanted and useless plants that grow along with the crop plants. Weeds compete with traps for light and nutrients, besides harbouring various pathogens. It is estimated that the world's crop yield is reduced by 10-15% due to the presence of weeds.

To tackle the problem of weeds, modern agriculture has developed a wide range of weed killers which are collectively referred to as herbicides. In general, majority of the herbicides are broad-spectrum as they can kill a wide range of weeds.

A good or an ideal herbicide is expected to possess the following characteristics:

1. Capable of killing weeds without affecting crop plants.

2. Not toxic to animals and microorganisms.

3. Rapidly trans-located within the target plant.

4. Rapidly degraded in the soil.

None of the commercially available herbicides fulfills all the above criteria. The major limitation of the herbicides is that they cannot discriminate weeds from crop plants. For this reason, the crops are also affected by herbicides, hence the need to develop herbicide-resistant plants. Thus, these plants provide an opportunity to effectively kill the weeds (by herbicides) without damaging the crop plants.

Strategies for Engineering Herbicide Resistance

A number of biological manipulations particularly involving genetic engineering are in use to develop herbicide-resistant plants.

1. Overexpression of the target protein: The target protein, being acted by the herbicide can be produced in large quantities so that the affect of the herbicide becomes insignificant. Overexpression can be achieved by integrating multiple copies of the genes and/or by using a strong promoter.

2. Improved plant detoxification: The plants do possess natural defense systems against toxic compounds (herbicides). Detoxification involves the conversion of toxic herbicide to non-toxic or less toxic compound. By enhancing the plant detoxification system, the impact of the herbicide can be reduced.

3. Detoxification of herbicide by using a foreign gene: By introducing a foreign gene into the crop plant, the herbicide can be effectively detoxified.

4. Mutation of the target protein: The target protein which is being affected by the herbicide can be suitably modified. The changed protein should be capable of discharging the functions of the native protein but is resistant to inhibition by the herbicide.

Once the resistant target protein gene is identified, it can be introduced into the plant genomes, and thus herbicide-resistant plants can be developed. For success in the development of herbicide resistant plants, good knowledge of the target protein and the action of herbicides is required.

Glyphosate Resistance

Glyphosate, is a glycine derivative. It acts as a broad-spectrum herbicide and is effective against 76 of the world's worst 78 weeds. Glyphosate is less toxic to animals and is rapidly degraded by microorganisms. In addition, it has a short half-life. The American chemical company Monsanto markets glyphosate as Round up.

Mechanism of Action of Glyphosate

Glyphosate is rapidly transported to the growing points of plants. It is capable of killing the plants even at a low concentration. Glyphosate acts as a competitive inhibitor of the enzyme 5-enoylpyruvylshikimate 3-phosphate synthase (EPSPS). This is a key enzyme in shikimic acid pathway that results in the formation of aromatic amino acids (tryptophan, phenylalanine and tyrosine), phenols and certain secondary metabolites.

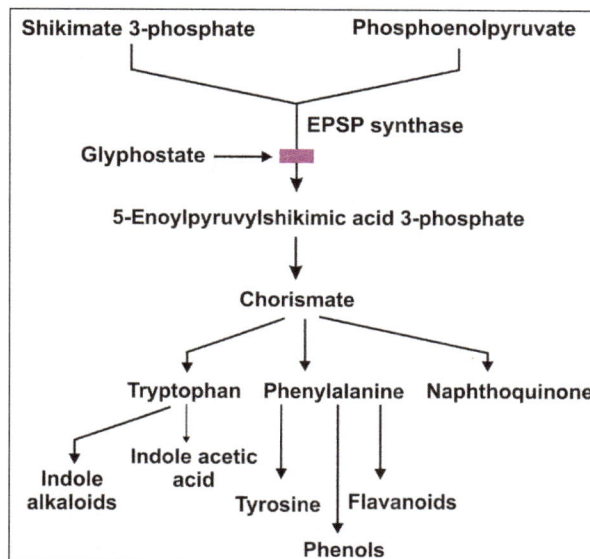

Shikimate pathway indicating the action of the herbicide glyphosate (EPSP synthase-5 -Enolylpyruvylshikimate 3- Phosphate synthase.

The enzyme EPSPS catalyses the synthesis of 5-enoylpyruvylshikimate 3-phosphate from shikimate 3-phosphate and phosphoenoylpyruvate. Glyphosate has some structural similarly with the substrate phosphoenol pyruvate. Consequently, glyphosate binds more tightly with EPSPS and blocks the normal shikimic acid pathway. Thus, the herbicide glyphosate inhibits the biosynthesis of aromatic amino acids and other important products.

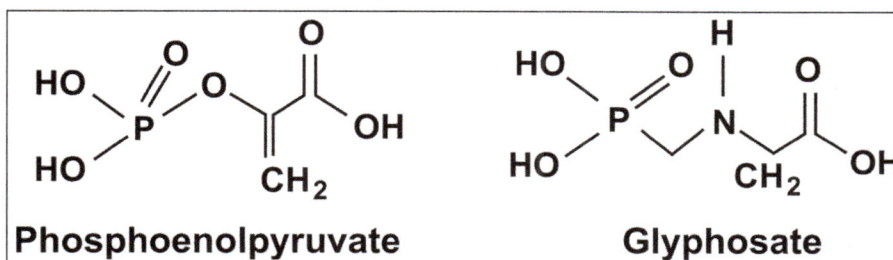

Structures of Phosphoenolpyruvate (the substrate) and the herbicide glyphosate (the competitive inhibitor).

This results in inhibition of protein biosynthesis (due to lack of aromatic amino acids). As a consequence, cell division and plant growth are blocked. Further, the plant growth regulator indole acetic acid (an auxin) is also produced from tryptophan. The net result of glyphosate is the death of the plants. Glyphosate is toxic to microorganisms as they also possess shikimate pathway.

Glyphosate is non-toxic to animals (including humans), since they do not possess shikimate pathway. Of the three aromatic amino acids (synthesized in this pathway), tryptophan and phenylalanine are essential and they have to be supplied in the diet, while tyrosine can be formed from phenylalanine.

Strategies for Glyphosate Resistance

There are three distinct strategies to provide glyphosphate resistance to plants:

1. Overexpression of crop plant EPSPS gene:

An overexpressing gene of EPSPS was detected in Petunia. This expression was found to be due to gene amplification rather than an increased expression of the gene. EPSPS gene from Petunia was isolated and introduced into other plants. The increased synthesis of EPSPS (by about 40 fold) in transgenic plants provides resistance to glyphosate. These plants can tolerate glyphosate at a dose of 2-4 times higher than that required to kill wild-type plants.

2. Use of mutant EPSPS genes:

An EPSPS mutant gene that conferred resistance to glyphosate was first detected in the bacterium Salmonella typhimurium. It was found that a single base substitution (C to 7) resulted in the change of an amino acid from proline to serine in EPSPS. This modified enzyme cannot bind to glyphosate, and thus provides resistance.

The mutant EPSPS gene was introduced into tobacco plants using Agrobacterium Ti plasmid vectors. The transgene produced high quantities of the enzyme EPSPS. However, the transformed tobacco plants provided only marginal resistance to glyphosate. The reason for this was not immediately identified.

It was later known that the shikimate pathway occurs in the chloroplasts while the glyphosate resistant EPSPS was produced only in the cytoplasm. This enzyme was not transported to the chloroplasts, hence the problem to provide resistance. This episode made scientists to realize the importance of chloroplasts in genetic engineering.

In later years, the mutant EPSPS gene was tagged with a chloroplast-specific transit peptide sequence. By this approach, the glyphosate-resistant EPSPS enzyme was directed to freely enter chloroplast and confer resistance against the herbicide.

3. Detoxification of glyphosate:

The soil microorganisms possess the enzyme glyphosate oxidase that converts glyphosate to glyoxylate and aminomethylphosponic acid. The gene encoding glyphosate oxidase has been isolated from a soil organism Ochrobactrum anthropi. With suitable modifications, this gene was introduced into crop plants e.g. oilseed rape. The transgenic plants were found to exhibit very good glyphosate resistance in the field.

Use of a Combined Strategy

More efficient resistance of plants against glyphosate can be provided by employing a combined strategy. Thus, resistant (i.e. mutant) EPSPS gene in combination with glyphosate oxidase gene are used. By this approach, there occurs glyphosate resistance (due to mutant EPSPS gene) as well as its detoxification (due to glyphosate oxidase gene).

Phosphinothricin Resistance

Phosphinothricin (or glufosinate) is also a broad spectrum herbicide like glyphosate. Phosphinothricin is more effective against broad-leafed weeds but least effective against perennials. Phosphinothricin and glufosinate are two names for the same herbicide. However, to avoid confusion between glyphosate and glufosinate, phosphinothricin is more commonly used. Basta Aventis and Liberty are the trade names for phosphinothricin.

Phosphinothricin-A Natural Herbicide

Phosphinothricin is an unusual herbicide, being a derivative of a natural product namely bialaphos. Certain species of Streptomyces produce bialaphos which is a combination of phosphinothricin bound to two alanine residues, forming a tripeptide. By the action of a peptidase, bialaphos is converted to active phosphinothricin.

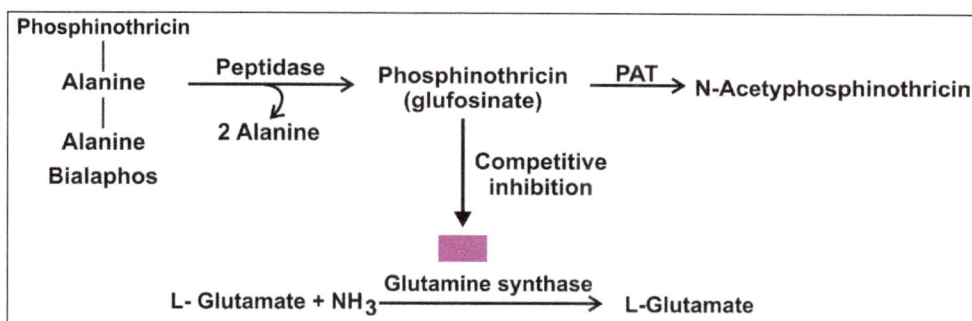

The Formation, Mode of Action and Detoxification of Phosphinothricin
(PAT-Phosphinothricin acetyl transferase).

Mechanism of Action of Phosphinothricin

Phosphinothricin acts as a competitive inhibitor of the enzyme glutamine synthase. This is possible since phosphinothricin has some structural similarity with the substrate glutamate. As a consequence of the inhibition of glutamine synthase, ammonia accumulates and kills the plant cells. Further, disturbance in glutamine synthesis also inhibits photosynthesis. Thus, the herbicidal activity of phosphinothricin is due to the combined effects of ammonia toxicity and inhibition of photosynthesis.

Strategy for Phosphinothricin Resistance

The natural detoxifying mechanism of phosphinothricin observed in Streptomyces sp has prompted scientists to develop resistant plants against this herbicide. The enzyme phosphinothricin acetyl transferase (of Streptomyces sp) acetylates phosphinothricin, and thus inactivates the herbicide.

The gene responsible for coding phosphinothricin acetyl transferase (bar gene) has been identified in Streptomyces hygroscopicus. Some success has been reported in developing transgenic maize and oilseed rape by introducing bar gene. These plants were found to provide resistance to phosphinothricin.

Sulfonylureas and Imidazolinones Resistance

The herbicides namely sulfonylureas and imidazolinones inhibit the enzyme acetolactate synthase (ALS), a key enzyme in the synthesis of branched chain amino acids namely isoleucine, leucine and valine. Mutant forms of this enzyme and the corresponding genes have been isolated, identified and characterized. Transgenic plants with the mutant genes of ALS were found to be resistant to sulfonylureas and imidazolinones e.g. maize, tomato, sugar beet.

Resistance to Other Herbicides

Besides the above, some other herbicide resistant plants have also been developed e.g. bromoxynil, atrazine, phenocarboxylic acids, cyanamide. A list of selected examples of gene transferred herbicide resistant plants is given.

Selected examples of gene transferrad herbicide resistant plants.

Herbicide	Genetransfer/Mechanism of resistance	Transgenic crop(s)
Glyphosate	Inhibition of EPSPS	Soybean, tomato
Glyphosate	Detoxification by glyphosate oxidase	Maize, soybean
Phosphinothricin	Bar gene coding phosphinothricin acetyltransferase	Maize, rice, wheat, cotton, potato, tomato, sugarbeet
Sulfonylureas imidazolinones	Mutant plant with acetolactate Synthase	Rice, tomato, maize, sugarbeet
Bromoxynil	Nitrilase detoxification	Cotton, potato, tomato
Atrazine	Mutant plant with chloroplast psb A gene	soybean
Phenocarboxylic acids	Monooxygenase detoxification(e.g. 2,4.D and 2.4.5-T)	Maize cotton
cyanamide	Cyanamide hydratase gene	Tabacco

It may however, be noted that some of the herbicide-resistant transgenic plants are at field-trial stage. Due to environmental concern, a few of these plants are withdrawn e.g. atrazine- resistant crops.

Environmental Impact of Herbicide-resistant Crops

The development genetically modified (GM) herbicide-resistant crops has undoubtedly contributed to increase in the yield of crops. For this reason, farmers particularly in the developed countries (e.g. USA) have started using these GM crops. Thus, the proportion of herbicide resistant soybean plants grown in USA increased from 17% in 1997 to 68% in 2001.

The farmer is immensely benefited as there is a reduction in the cost of herbicide usage. It is believed that the impact of herbicide-resistant plants on the environment is much lower than the direct use of the herbicides in huge quantities.

There are however, other Environmental Concerns:

i.   Disturbance in biodiversity due to elimination of weeds.

ii. Rapid development of herbicide-resistance weeds that may finally lead to the production of super weeds.

Tolerance to Water Deficit Stresses

The environmental conditions such as temperature (heat, freezing, chilling), water availability (shortage due to drought), and salinity influence the plant growth, development and yield. The abiotic stresses due to temperature, drought and salinity are collectively regarded as water deficit stresses.

Causes of Water Deficit

Water deficit may occur due to the following causes:

i.   Reduced soil water potential.

ii. Increased water evaporation (in dry, hot and windy conditions).

iii. High salt concentration in the soil (decrease soil water potential).

iv. Low temperature resulting in the formation of ice crystals.

Effects of Water Deficit

Results in osmotic stress.

Inhibits photosynthesis.

iii. Increases the concentration of toxic ions (reactive oxygen species) within the cells.

iv. Loss of water from the cell causing plasmolysis and finally cell death.

Tolerance to Osmotic Stress

The plant cells are subjected to severe osmotic stress due to water deficit. They however, produce certain compounds, collectively referred to as osmoprotectants or osmolytes, to overcome the osmotic stress. Osmoprotectants are non-toxic compatible solutes and are divided into two groups.

i.   Sugar and sugar alcohols e.g. mannitol, sorbitol, pinitol, ononitol, trehalose, fructans.

ii. Zwitterionic compounds: These osmoprotectants carry positive and negative charges e.g. proline, glycine betaine.

The production of a given osmoprotectant is species dependent. The formation of mannitol, proline and glycine betaine are more closely linked to osmotic tolerance.

Strategies to Develop Water Deficit Tolerance Plants

As explained above osmoprotectants offer good protection to plants against osmotic stress and therefore water deficit. It is therefore, logical to think of genetic engineering strategies for the increased production of osmoprotectants.

Some progress has been made in this direction. The biosynthetic pathways for the production of many osmoprotectants have been established and genes coding key enzymes isolated. In fact, some progress has been made in the development of transgenic plants with high production of osmoprotectants.

Transgenic Plants with Glycine Toetaine Production

Glycine betaine is a quaternary ammonium compound and is electrically neutral. Besides functioning as a cellular osmolytes, glycine betaine stabilizes proteins and membrane structures. Some of the key enzymes for the production of glycine betaine have been identified e.g. choline mono-oxygenase, choline dehydrogenase, betaine aldehyde dehydrogenase.

The genes coding these enzymes were transferred to develop transgenic plants. By using choline oxidase gene from Arthrobacter sp, transgenic rice that produces higher glycine betaine (which offers tolerance against water deficit stress) has been developed.

Resistance against Ice-nucleating Bacteria

Formation of ice on the plant cells (outer membrane) is a complex chemical process. The importance of ice-nucleating bacteria is recognized in recent years. The occurrence of these bacteria has been reported in most of the plants — cereals, fruits and vegetable crops. The ice-nucleating bacteria synthesize proteins, which coalesce with water molecules to form ice crystals at temperature around 32 °F. As the ice crystals grow, they can pierce the plant cells and severely damage the plants.

Chemical Treatment of Plants to Protect from Ice Formation

Plants can be treated with copper containing compounds to kill the bacteria. Another approach is to use urea solution so that the ice formation is minimized.

Ice-Minus Bacteria to Resist Plants from Cold Temperatures

The bacterium Pseudomonas syringae is one of the highly prevalent ice-forming organisms in nature. With genetic manipulations, the gene that directs the synthesis of ice-related bacterial proteins in P. syringae was removed. These newly developed bacteria are referred to as ice-minus bacteria.

The researchers proposed to spray the transgenic ice-minus bacteria on to young plants. The intention was that these bacteria would give frost tolerance to the plants; and thus increase the crop yield. The opponents of DNA technology were against this approach—the main fear being that the bacterial mutants may create some health complication in humans.

The researchers argued and justified that no new genetic information is introduced into P. syringae and it is closely related in all aspect to the parent one which is already in the environment. After prolonged court proceedings in USA, clearance was given for spraying the ice-minus bacteria in the fields.

It was in 1987, ice-minus bacteria were sprayed on to the field of potato plants and strawberry plants. Another strain of P. syringae commercially labeled, as Frostban was later developed and used in crop fields. It may be noted here that ice-minus bacteria of P. syringae were the first transgenic bacteria that were used outside the laboratory. Fortunately, the experiments yielded encouraging results, since crop damage due to frost formation was found to be reduced.

Arabidopsis with Cold-tolerant Genes

Scientists were successful in developing cold- tolerant genes (around 20) in Arabidopsis when this plant was gradually exposed to slowly declining temperatures. They also identified a coordinating gene that encodes a protein, which acts as a transcription factor for regulating the expression of cold-tolerant genes. By introducing the coordinating gene, expression of cold-tolerant genes was triggered, and this protected the plants against cold temperatures. More work is in progress in this direction.

Improvement of Crop Yield and Quality

With the advances made in plant genetic engineering, improvement in crop yield and quality have become a reality. The crop yield is primarily dependent on the photosynthetic efficiency and the harvest index (the fraction of the dry matter allocated to the harvested part of the crop). The quality of the crop is dependent on a wide range of desirable characters-nutritional composition of edible parts, flavour, processing quality, shelf-life etc.

Green Revolution

The 'Green Revolution' led by Borlaug, Swaminathan and Khus enabled the world's food supply to be tripled during the last three decades of 20th century. This was made possible by adopting genetically improved varieties of crops, coupled with advances in crop management.

The development of high-yielding varieties of wheat and rice has enabled several developing countries (a good example being India) to move from a position of food scarcity to become net exporter of these cereals.

The Green Revolution became a reality as the farmers adopted to new cereal seeds, besides employing high-input methods of agriculture — use of nitrogen fertilizers, herbicides, pesticides, modern equipment of agriculture etc.

Selected Examples of Crops for Quality and Yield:

There are a wide range of crops that have been manipulated by scientists for improved yield and quality. Only selected examples are briefly described.

Genetic Engineering for Extended Shelf-life of Fruits

The genetic manipulation of fruit ripening has become an important commercial aspect in plant genetic engineering.

Delay in Fruit Ripening has many Advantages:

i. It extends the shelf-life, keeping the quality of the fruit intact.

ii Long distance transport becomes easy without damage to fruit.

iii. Slow ripening improves the flavour.

Genetic engineering work has been extensively carried out in tomatoes, and some of the development are described.

Biochemical Changes During Tomato Ripening

Fruit ripening is an active process. It is characterized by increased respiration accompanied by a rapid increase in ethylene synthesis. As the chlorophyll gets degraded, the green colour of the fruit disappears, and a red pigment, lycopene is synthesized.

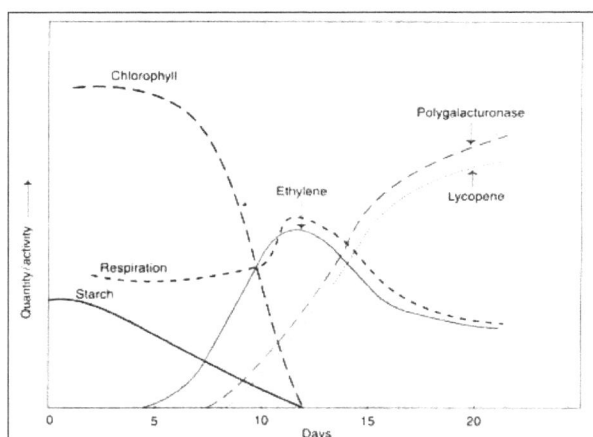

Biochemical Changes during the Process of Tomato Ripening.

The fruit gets softened as a result of the activity of cell wall degrading enzymes namely polygalacturonase (PG) and pectin methyl esterase. The phytohormone ethylene production is intimately linked to fruit ripening as it triggers the ripening process of fruit. Addition of exogenous ethylene promotes fruit ripening, while inhibition of ethylene biosynthesis drastically reduces ripening.

The breakdown of starch to sugars, and accumulation of a large number of secondary products improves the flavour, taste and smell of the fruit. Three distinct genes involved in tomato ripening have been isolated and cloned. The enzymes encoded by these genes and their respective role in fruit ripening are given in table.

Clones of tomato ripening genes and their functions.

Geme cone	Enzyme synthesized	Function in ripening
pTOM 5	Phytoene synthase	Lycopene synthesis that gives red colouration
pTOM 6	Polygalacturonase	Degradation of cell wall, resulting in fruit softening
pTOM 13	ACC oxidase	Ethylene formation that triggers fruit ripening

Genetic Manipulations of Fruit Ripening

Scientists have been trying to genetically manipulate and delay the fruit ripening process. Almost all the attempts involve antisense RNA approach. Manipulation of the enzyme polygalacturonase (development of Flavr Savr tomato):

As already stated, softening of the fruit is largely due to degradation of the cell wall (pectin) by the enzyme polygalacturonase (PG). The gene responsible for PG, the rotting enzyme, has been cloned (pTOM 6). The genetic manipulation of polygalacturonase by antisense RNA approach for the development of Flavr Savr tomato (by Calgene Company in USA) is depicted in figure and mainly involves the following stages.

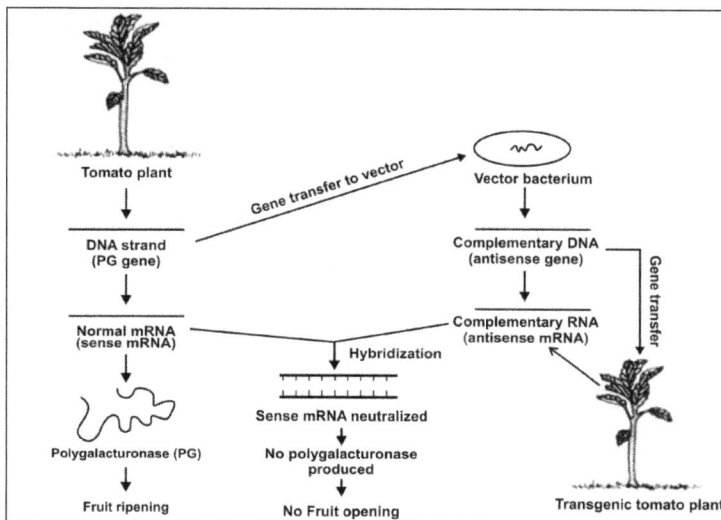

Genetic Manipulation of the Enzyme Polygalacturonase (PG) by antisense RNA approach (Production of Flavr Savr tomato plant).

1. Isolation of the DNA from tomato plant that encodes the enzyme polygalacturonase (PG).

2. Transfer of PG gene to a vector bacteria and production of complementary DNA molecules.

3. Introduction of complementary DNA into a fresh tomato plant to produce a transgenic plant.

Mechanism of PG antisense RNa Approach

In the normal tomato plant, PG gene encodes a normal (sense) mRNA that produces the enzyme polygalacturonase that is actively involved in fruit ripening. The complementary DNA of PG encodes for antisense mRNA, which is complementary to normal (sense) mRNA. The hybridization between the sense and antisense mRNAs renders the sense mRNA ineffective. Consequently, no polygalacturonase is produced, hence fruit ripening is delayed.

The Rise and Fall of Flavr Savr Tomato

The genetically engineered tomato, known as Flavr Savr (pronounced flavour saver) by employing PC antisense RNA was approved by U.S. Food and Drug Administration on 18th May 1994. The FDA ruled that Flavr Savr tomatoes are as safe as tomatoes that are bred by conventional means, and therefore no special labeling is required. The new tomato could be shipped without refrigeration too far off places, as it was capable of resisting rot for more than three weeks (double the time of a conventional tomato).

Although Flavr Savr was launched with a great fanfare in 1995, it did not fulfill the expectation for the following reasons:

1. Transgenic tomatoes could not be grown properly in different parts of U.S.A.

2. The yield of tomatoes was low.

3. The cost of Flavr Savr was high.

It is argued that the company that developed Flavr Savr, in its overenthusiasm to become the first Biotech Company to market a bioengineered food had not taken adequate care in developing the transgenic plant. And unfortunately, within a year after its entry, Flavr Savr was withdrawn, and it is now almost forgotten.

Manipulation of Ethylene Biosynthesis

It has been clearly established that ethylene plays a key role in the ripening of fruits. The biosynthetic pathway of ethylene is depicted. Ethylene is synthesized from S-adenosyl methonine via the formation of an intermediate, namely 1-aminocyclopropane-1 carboxylic acid (ACC), catalysed by the enzyme ACC synthase. The next step is the conversion of ACC to ethylene by ACC oxidase.

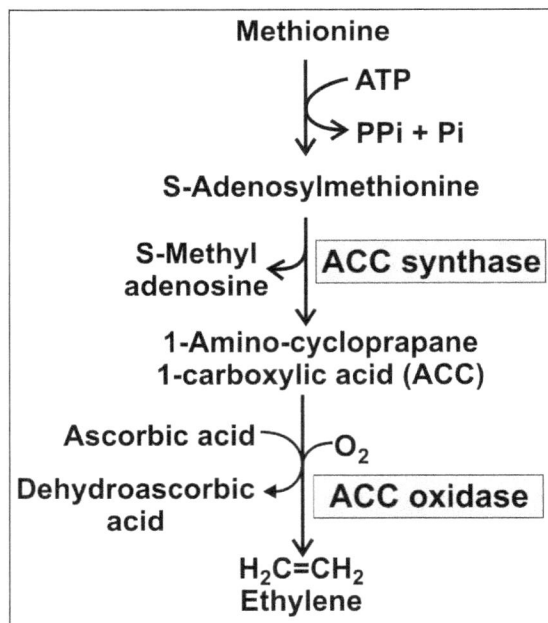

Methionine
$\quad$ ATP
$\quad$ PPi + Pi
S-Adenosylmethionine
S-Methyl adenosine $\leftarrow$ | ACC synthase |
1-Amino-cycloprapane 1-carboxylic acid (ACC)
Ascorbic acid $\quad O_2$
Dehydroascorbic acid $\leftarrow$ | ACC oxidase |
$H_2C=CH_2$
Ethylene

Biosynthesis of Ethylene

Three different strategies have been developed to block ethylene biosynthesis, and thereby reduce fruit ripening.

1. Antisense gene of ACC oxidase: Transgenic plants with antisense gene of ACC oxidase have been developed. In these plants, production of ethylene was reduced by about 97% with a significant delay in fruit ripening.

2. Antisense gene of ACC synthase: Ethylene biosynthesis was inhibited to an extent of 99.5% by inserting antisense gene of ACC synthase, and the tomato ripening was markedly delayed.

3. Insertion of ACC deaminase gene: ACC deaminase is a bacterial enzyme. It acts on ACC (removes amino group), and consequently the substrate availability for ethylene biosynthesis is reduced. The bacterial gene encoding ACC deaminase has been transferred and expressed in tomato plants. These transgenic plants inhibited about 90% of ethylene biosynthesis. The fruit ripening was delayed by about six weeks. The strategies 1 and 2 may be referred to as antisense ethylene technology.

Longer Shelf-life of Fruits and Vegetables

The spoilage of fruits, vegetables and senescence of picked flowers, collectively referred to as post-harvest spoilage is major concern in agriculture. This hampers the distribution system particularly when the transport is done to far off places. The successful manipulations to delay ripening, senescence and spoilage of various foods will significantly contribute to the appropriate food distribution and thus good economic practices in agriculture.

Suppressing the biosynthesis of ethylene appears to be a promising area to reduce the spoilage of fruits, vegetables and senescence of flowers. The three different strategies to block ethylene synthesis in tomato have been described. The same approaches in fact can be successfully used for other fruits, vegetables etc., to achieve longer shelf- life.

Genetic Engineering for Preventing Discoloration

Discoloration of fruits and vegetables is a major postharvest problem encountered in food industry. Certain food additives are added to prevent discoloration. However, these additives may cause health complications in humans.

Biochemically, discoloration of fruits and vegetables is mainly due to the oxidation of phenols (mono- and diphenols) to quinones, catalysed by a group of enzymes namely polyphenol oxidases. These enzymes are localized in the membranes of mitochondria and chloroplasts. Genetic manipulations using antisense approach to inhibit the synthesis of polyphenol oxidase has been carried. Some success has been reported in preventing the discoloration of potatoes by this strategy.

Genetic Engineering for Flower Pigmentation

There are continuous attempts in flower industry to make the ornamental flowers more attractive (by improving or creating new colours), besides prolonging post-harvest lifetime. The cut flower industry is mostly (about 70%) dominated by four plants—roses, tulips, chrysanthemums and carnations.

The most common type of flower pigments are anthocyanin's, a group of flavonoids. They are synthesized by a series of reactions, starting from the amino acid phenylalanine. The colour of the flower is dependent on the chemical nature of the anthocyanin produced.

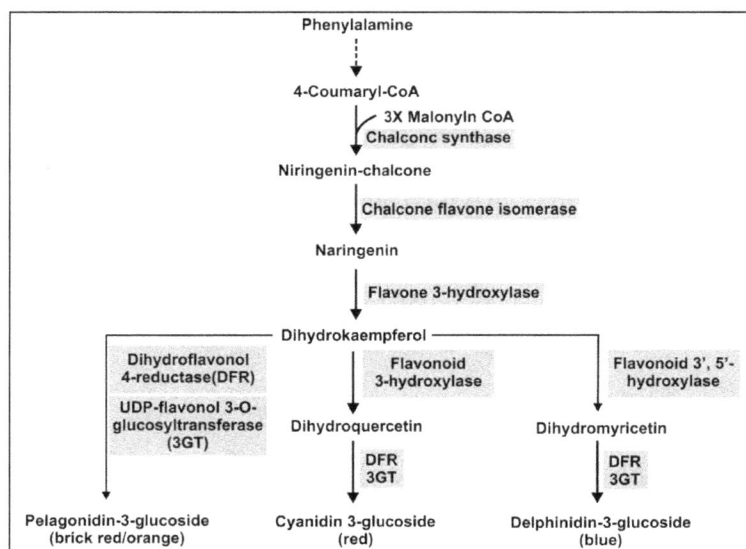

Biosynthesis of Anthocyanins

i. Pelagonidin 3-glucoside — brick red/orange.

ii. Cyanidin 3-glucoside — red.

iii. Delphinidin 3-glucoside — blue to purple.

Manipulation of Anthocyanin Pathway Enzymes

The enzymes responsible for different reactions, in the anthocyanin pathway have been identified. By genetic manipulations and mutations, it is possible to develop flowers with the desired colours. Most of the flowers (roses, carnations chrysanthemums) lack blue colour due to the absence of the key enzyme flavonoids 3', 5'- hydroxylase (F 3' 5' H) that produces delphinidine 3 glucoside. One company, by the name Florigene, has genetically manipulated and introduced the gene encoding the enzyme F 3' 5' H (from Petunia hybrida) into the following plants.

The world's first genetically modified (GM) flower was introduced in 1996. It was a mauve (bluish) coloured carnation with a trade name Moondust. Subsequently, many other flowers have been produced and marketed.

Can GM-flowers be Eaten

In majority of countries, flowers are used for ornamental purposes and not usually eaten. However, in some countries like Japan, flower petals are used for decoration of foods, and frequently eaten also. This raises an important question about the safety of GM flowers, since they are not thoroughly screened for human consumption. But the present belief is that the anthocyanin's (the colouring chemical molecules) are natural plant materials, and their consumption may be in fact beneficial to health.

Genetic Engineering for Male Sterility

The plants may inherit male sterility either from the nucleus or cytoplasm, cytoplasmic male sterility (cms) is due to the defects in the mitochondrial genome. It is possible to introduce male sterility through genetic manipulations while the female plants maintain fertility.

In tobacco plants, male sterility was introduced by using a mitochondrial mutated gene encoding the enzyme ribonuclease. The gene encoding ribonuclease namely barnase gene from Bacillus amyloliquefaciens was transferred to tobacco plants.

The ribonuclease is toxic to tapetal cells, and thus prevents the development of pollen, ultimately leading to male sterility. By this approach, transgenic plants of tobacco, cauliflower, cotton, tomato, corn, lettuce etc., with male sterility have been developed. It is possible to restore male sterility in the above plants by crossing them with a second set of transgenic plants containing ribonuclease inhibitor gene.

Transgenic Plants with Improved Nutrition

Genetic manipulations for improving the nutritional quality of plant products are of great importance in plant biotechnology. Some success has been achieved in this direction through conventional cross-breeding of plants. However, this approach is very slow and difficult, and many a times will not give the traits with the desired improvements in the nutritional quality. Selected examples of genetic engineering with improved nutritional contents are described.

Amino Acids of Seed Storage Proteins

Of the 20 amino acids present in the humans, 10 are essential while the other 10 can be synthesized by the body. The 10 essential amino acids (EAAs) have to be supplied through the diet. Cereals (rice, wheat, maize, corn) are the predominant suppliers of EAAs. However, cereals do not contain adequate quantity of the essential amino acid lysine.

On the other hand, pulses (Bengal gram, red gram, soybean) are rich in lysine and limited in sulfur-containing amino acids (the essential one being methionine). Transgenic routes have been developed to improve the essential amino acid contents in the seed storage proteins of various crop plants.

Overproduction of Lysine by Deregulation

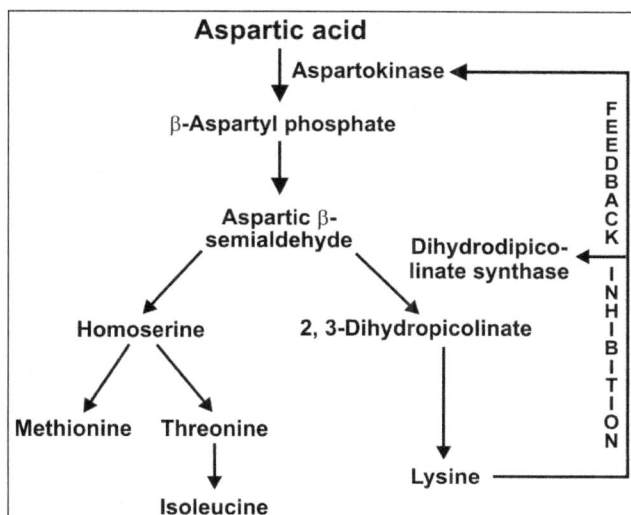

An overview of the Biosynthetic Pathway of Some Essential Amino Acids derived from aspartic acid.

The four essential amino acids namely lysine, methionine, threonine and isoleucine are produced from a non-essential amino acid aspartic acid. The formation of lysine is regulated by feedback inhibition of the enzymes aspartokinase (AK) and dihydrodipicolinate synthase (DHDPS). Theoretically, it is possible to overproduce lysine by abolishing the feedback regulation. This is what has been accomplished.

The lysine feedback-insensitive genes encoding the enzymes AK and DHPDS have been respectively isolated from E.coli and Cornynebacterium with appropriate genetic manipulations, these genes were introduced into soybean and canola plants. The transgenic plants so developed produced high quantities of lysine.

Transfer of Genes Encoding Methionine-rich Proteins

Several genes encoding methonine-rich proteins have been identified:

    i.   In maize, 21 KDa zein with 28% methionine.

    ii.   In rice, 10 KDa prolamin with 20% methionine.

    iii.   In sunflower, seed albumin with 16% methionine.

These genes have been introduced into some crops such as soybean, maize and canola.

The transgenic plants produced proteins with high contents of sulfur-containing amino acids.

Production of Lysine-rich Glycinin in Rice

Glycinin is a lysine-rich protein of soybean. The gene encoding glycinin has been introduced into rice and successfully expressed. The transgenic rice plants produced glycinin with high contents of lysine. Another added advantage of glycinin is that its consumption in humans is associated with a reduction in serum cholesterol (hypo- cholesterolaemic effect).

Construction of Artificial Genes to Produce Proteins Rich in EAAs

Attempts are being made to construct artificial genes that code for proteins containing the essential amino acids in the desired proportion. Some success has been reported in the production of one synthetic protein containing 13% methionine residues.

Genetic Engineering for Improving Palatability of Foods

More than the nutritive value, taste of the food is important for attracting humans. It is customary to make food palatable by adding salt, sugar, flavors and many other ingredients. It would be nice if a food has an intrinsically appetizing character.

A protein monellin isolated from an African plant (Dioscorephyllum cumminsii) is about 100,000 sweeter than sucrose on molar basis. Monellin gene has been introduced into tomato and lettuce plants. Some success has been reported in the production of monellin in these plants, improving the palatability.

Golden Rice—The Provitamin a Enriched Rice

About one-third of the world's population is dependent on rice as staple food. The milled rice that is usually consumed is almost deficit in P-carotene, the pro-vitamin A. As such, vitamin A deficiency (causing night blindness) is major nutritional disorder world over, particularly in people subsisting on rice.

To overcome vitamin A deficiency, it was proposed to genetically manipulate rice to produce β-carotene, in the rice endosperm. The presence of β-carotene in the rice gives a characteristic yellow/orange colour, hence the pro-vitamin A-enriched rice are appropriately considered as Golden Rice.

In figure an outline of the biosynthetic pathway for the formation of P-carotene is given. The genetic manipulation to produce Golden Rice required the introduction of three genes encoding the enzymes phytoene synthase, carotene desaturase and lycopene β-cyclase. It took about 7 years to insert three genes for developing Golden Rice.

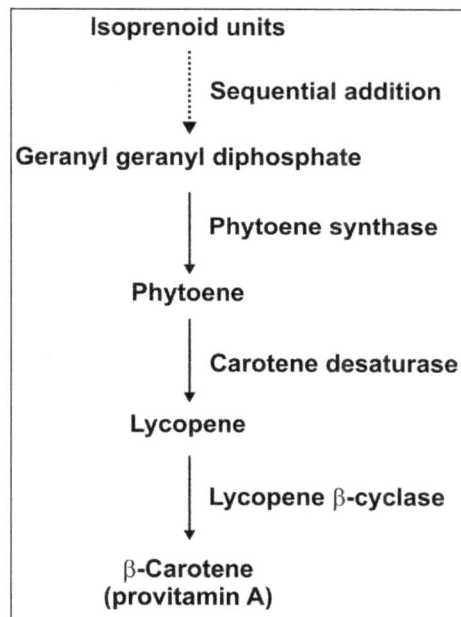

An outine of Pathway for the Biosynthesis of Provitamin A (β-carotene).

Golden Rice has met almost all the objections raised by the opponents of GM foods. However, many people are still against the large scale production of Golden Rice, as this will open door to the entry of many other GM foods.

Another argument put forth against the consumption of Golden Rice is that it can supply only about 20% of daily requirement of vitamin A. But the proponents justify that since rice is a part of a mixed diet consumed (along with many other foods), the contribution of pro-vitamin A through Golden Rice is quite substantial.

Recently, a group of British scientists have developed an improved version of Golden Rice. The new strain, Golden Rice 2 contains more than 20 times the amount of pro-vitamin A than its predecessor. It is claimed that a daily consumption of 70 g rice can meet the recommended dietary allowance for vitamin A.

Genetic Engineering to Increase Vitamins and Minerals

The transgenic rice (Golden Rice) developed with high pro-vitamin A content is described above. Transgenic crop plants are also being developed for increased production of other vitamins and minerals. A transgenic Arabidopsis thaliana that can produce ten-fold higher vitamin E (α-tocopherol) than the native plant has been developed. This was done by a novel approach. A. thaliana possesses the biochemical machinery to produce a compound close in structure to α-tocopherol.

A gene that can finally produce α-tocopherol is also present, but is not expressed. This dormant gene was activated by inserting a regulatory gene from a bacterium. This resulted in an efficient production of vitamin E. Some workers are trying to increase the mineral contents of edible plants by enhancing their ability to absorb from the soil. Some success has been reported with regard to increased concentration of iron.

Commercial Transgenic Crop Plants

The very purpose of production of transgenic plants is for their commercial importance with high productivity. It was in 1995-96; transgenic plants (potato and cotton) were, for the first time, made available to farmers in USA. By the year 1998-99, five major transgenic crops (cotton, maize, soybean, canola and potato) were in widespread use. They accounted for about 75% of the total area planted by crops in USA.

A selected list of transgenic crop plants (approved in USA) for the commercial use is given in. Some examples of transgenic crop plants, which are at the developmental stages are given. These plants are carefully designed to give rise to products, which will improve human health and increase of crop yield.

A seletd list of transgenic crop plants (GM crops approved in USA) for commercial use		
Crop plant	Genetically altered trait	Product name
Cotton	Insect resistance	Bollguard
	Glayphosale resistance	Roundup ready
	Bromoxynil resistance	BXN
	Sulfonylurea resistance	
Maize	Insect resistance	Yield guard
	Insect resistance	Maximize
	Glyphosate resistance	Roundup Ready
	Glufosinate resistance	Liberty link
Rice	Vitamin A enrichment	Golden Rice
Tomato	Delayed ripening	Flavr savr
	Delayed ripening	Endless Summer
	Virus reisitance	
Soybean	Glyphosate resistance	Roundup Ready
potato	Insect resistance	Newleaf
	Modified starch	
Oilseed rape (canola)	Glufosinate resistance	Innovator

	Glyphosate resistance	Roundup Ready
	High lauric acid	Laurical
	Male sterility hybrid	
Squesh	Virus resisstance	Freedom ll
Tobacco	Virus resisstance	
Capsicum	Virus resisstance	
Carnation	Modified flower colour	

Some examples of transgenic crop plants (gm plants) at the developmental stages		
plant	Gene transfer	Trait transferred/application(s)
For improving human health		
Tomato	Phytoene desaturase	Provitamin A (β-carotene) supplement
Canola	Υ-Tocopherol methyl transferase	Vitamin E supplement
Sugar beet	sucrose-sucrose fructosyl transferase	Fructans-low calorie alternatives to sucrose
Rice	Ferritin	Iron supplement
Potato	Antisense threonine synthase	Increased methionine levels
Potato	Seed albumin	Protein with all essential amino acids
Tomato	S-Adenosylmethionine decarboxy-lase	Increased lycopene levels
Tomato	Chalcone isomerase	Flavanols-act as antioxidants.
		Reduce risl of cancer, heart diseases
Arabidopsis	Isoflavone synthase	Isoflavones-reduce serum cholesterol, and reduce osteoporosis
Canola	Modified acyl-acyl carrier protein thioesterase	Cis-Stearates-lower the risk of heart diseases
For increased crop yield		
Rice	Phosphoenol pyruvate carboxylase	Increased efficiency of photosynthesis
Tobacco	Phytochrome A	Avoids shades
Lettuce	Gibberellic acid (GA) oxidase	Inhibits GA accumulation and stem growth (dwarfing)
Potato	Phytochrome B	Increased photosynthesis and longer life span
Others		
Tobacco and soybean	Cytochrome P$_{450}$	Synthesis of epoxy fatty acids for manufacture of adhesives and paints
Rice	Nicotianamine aminotransferase	tolerance to low iron availability
Tobacco	Nitroreductase	Reduces land contamination by trinitrotoluene

Goals of Biotechnological Improvements in Crops

There are about 30-40 crops that have been genetically modified, and many more are being added. However, very few of them have got the clearance for commercial use. A selected list is already given.

The ultimate goals of genetically modified (CM) crop plants are listed below:

   i.   Resistance to diseases (insect, microorganisms).

   ii.   Improved nitrogen fixing ability.

   iii.   Higher yielding capacity.

   iv.   Resistance to drought and soil salinity.

   v.   Better nutritional properties.

   vi.   Improved storage qualities.

   vii.   Production of pharmaceutically important compounds.

   viii.   Absence of allergens.

   ix.   Modified sensory attributes e.g. increased sweetness as in thaumatin.

Concerns about Transgenic plants

The fears about the harmful environmental and hazardous health effects of transgenic plants still exist, despite the fact that there have been no reports so far in this regard. The transfer of almost all the transgenic plants from the laboratory to the crop fields is invariably associated with legal and regulatory hurdles, besides the social and economic concerns.

The major concern expressed by public (also acknowledged by biotechnologists) is the development of resistance genes in insects, generation of super weeds etc. Several remedial measures are advocated to overcome these problems.

The farmers in developing countries are much worried about the seed terminator technology which forces them to buy seeds for every new crop. These farmers are traditionally habituated to use the seeds from the previous crop which is now not possible due to seed terminator technology.

Transgenic Plants as Bioreactors

Another important application of genetically transformed plants is their utility as bioreactors to produce a wide range of metabolic and industrial products.

Disease Resistance in Plants

Developing Virus Resistance Food Crops

Viruses are among the most ubiquitous pests in agriculture. Scientists are working to develop viral resistance in a variety of crops including squash, potato, sweet potato, wheat, papaya and raspberries.

Viruses are studied widely because they not only cause disease in humans, plants, animals and insects, but also are used as tools in the study of molecular biology and, in some cases, in the development of vaccines to fight the diseases they can cause.

Several techniques for virus resistance have been developed. These include viral coat protein technology and multiple gene transfers. A viral coat protein acts like a vaccine, causing the plant to develop resistance to the particular virus. Transferring the gene for a viral coat protein, a part of the outer shell of a virus that does not cause disease, into a plant acts like a vaccine for the plant.

The plant is then able to resist the virus, analogous to the way vaccines keep us from getting certain diseases like measles. The advantage of introducing only the coat protein is that it induces resistance without the introduction of the actual virus. The technique has been used successfully in many plants against several different viruses.

The first genetically engineered virus-resistant food crop in the marketplace was yellow crookneck squash. Using the viral coat protein approach, this squash was engineered to resist the watermelon mosaic virus and the zucchini yellow mosaic virus. Potatoes are highly susceptible to many viruses, including the potato mosaic virus and the potato leaf roll virus.

A leaf roll virus epidemic in 1996 was responsible for heavy potato crop losses in Idaho. The virus, spread by aphids, damaged the potatoes to the point that they were unmarketable. Scientists in Mexico, in collaboration with researchers at Monsanto, have developed potatoes resistant to several forms of this virus. Research on disease-resistant potatoes is continuing at other laboratories.

The feathery mottle virus has a damaging effect on sweet potatoes. In 1991, researchers began genetically engineering varieties of sweet potato grown in Africa, where it is an important subsistence crop. The sweet potato was engineered with coat protein from this virus and replicase genes. Replicase is an enzyme involved in the duplication of certain viral RNA molecules.

Current field-testing has demonstrated successful gene transformations and the desired development of resistance to sweet potato feathery mottle virus. Although wheat is an important food source, development of genetically engineered varieties has been slower than in corn, soy and cotton.

A major pest in wheat is barley yellow dwarf virus, which can cause damage in major wheat-growing regions such as North Dakota, because no resistant strains are known. Work is in progress to engineer resistance to this disease using the viral coat protein technique.

The wheat genome is highly complex-ten to twenty times larger than that of cotton or rice-and carries an exceptionally large amount of repetitive DNA sequences. Thus, targeting particular genes is challenging, and transgenic wheat biotechnology has advanced more slowly than that of other crops.

The papaya crop in Hawaii was nearly wiped out in the 1950s by the papaya ring-spot virus (PRSV). Transmitted by aphids, this virus causes one of the most serious diseases of papaya worldwide. Work to develop a transgenic virus-resistant variety began in the late 1980s. By 1992, resistant lines were field-tested; approvals for commercialization were granted in 1997.

The transgenic- resistant papaya is now in wide use in Hawaii, and similar work is in progress in the Philippines, Malaysia, Thailand, Vietnam and Indonesia to enhance resistance in local papaya varieties where ring-spot virus is a major pest. Researchers are also modifying other fruits for virus resistance.

Developing Fungi Resistance Food Crops

The search for genetic engineering tactics to combat fungi has intensified with the need to find adequate substitutes for fungicides such as methyl bromide, widely used on fruit and vegetables but being phased out due to its links to ozone depletion.

One emerging area is directed at a plant's production of defensins, a family of naturally occurring antimicrobial proteins which enhance the plant's tolerance to pathogens, especially bacteria. Certain defensins also demonstrate an ability to fight fungal infections. Defensins are found throughout nature in insects, mammals (including humans), crustaceans, fish and plants.

Defensins from moths and butterflies, the fruit fly, pea seeds and alfalfa seeds all show potent antifungal activity. The first transgenic application of defensins was the incorporation into potatoes of the antifungal defensin from alfalfa. Laboratory and field trials showed that the transgenic potatoes were as resistant to the fungal pathogen Verticillium dahliae as non-transgenic potatoes treated with fungicide.

Although studies are continuing, the chance that fungi will build resistance to defensins is thought unlikely. No known resistant strains of bacteria or fungi have yet evolved that can overcome these highly protective, pesticidal proteins.

On-going research involving banana and cassava is directed to cloning resistance genes for major tropical diseases such as black sigatoka, a leaf fungus that widely infects bananas, cassava mosaic disease and cassava bacterial blight.

In bananas, transgenic lines combining several antifungal genes have been generated. Selected lines are currently being tested for resistance to black sigatoka and Panama disease under greenhouse and field conditions.

Scientists are devising protection against the plant fungus Botrytis cinerea, a serious pathogen in wheat and barley. The strategy uses the gene for a natural plant defence compound named resveratrol. Scientists have also introduced a gene from a wine grape into barley to confer resistance to Botrytis cinerea. Field trials are underway.

Resistance to potato late blight, a disease caused by Phytophthora infestans, receives high priority in potato research. Plant disease from this fungus can be destructive to crop production, as was dramatically illustrated in the Irish potato famine.

In 1995, a U.S. late blight epidemic (caused by new aggressive strains of Phytophthora infestans) affected nearly 160,000 acres of potatoes, or about 20 per cent of domestic production.

Research is underway to genetically engineer potatoes that express the enzyme glucose oxidase and develop resistance to Phytophthora blights (Douches undated). At present, however, no products are close to commercialisation. Potatoes are also being transformed using a soybean gene for a protein (beta-1, 3-endoglucanase) that confers resistance to infection by Phytophthora.

Other studies report that transgenic potatoes expressing a protein called osmotin showed reduced damage from lesion growth in leaves inoculated with the Phytophthora infestans pathogen.

Still other research is attempting to boost fungal resistance in potatoes by transferring resistance genes from peas. Infection of these transgenic potatoes with the fungus triggers hormone-like signals in the potatoes that turn on the pea resistance genes.

One substance that is produced, chitosan, stops fungal growth and activates the potato's own natural defence systems. In rice, blast and sheath blight are major fungal diseases. Scientists created transgenic strains resistant to sheath blight that are currently being field-tested.

Developing Bacteria Resistance Food Crops

Most food crops are susceptible to bacterial diseases, but bacteria rarely attack certain plants, such as mosses, ferns and conifers. Bacterial infections in plants may cause leaf and fruit spots (lesions), soft rots, yellowing, wilting, stunting, tumours, scabs or blossom blights.

When tissue damage occurs on the blossoms, fruit or roots of food crops, yields may be reduced. Potatoes are susceptible to blackleg and soft rot diseases caused by the bacterial pathogen Erwinia carotovora.

To combat these bacteria, scientists have exploited the family of enzymes known as lysozymes that catalyze the breakdown of bacterial cell walls. Using cloned lysozyme genes and a promoter, transgenic potatoes were created that produced lysozyme.

In laboratory tests, the transformed potatoes exhibited substantially enhanced resistance to Erwinia carotovora. Field tests and further development of resistant lines are in progress. A different transgenic strategy to combat Erwinia carotovora was demonstrated in tobacco engineered to overexpress a peptide that kills bacteria.

The genetically engineered tobacco plants were resistant to both Erwinia carotovora and Pseudomonas syringae pv tabaci, the pathogen responsible for wild fire disease in rice. Scientists have also successfully transferred a bacterial resistance gene from wild rice to cultivated rice.

Developing Insects Resistance Food Crops

There are several different combat tactics, including engineering for the expression of toxins in plants that kill insects when they consume the plant material, but are nontoxic to other species that eat the plant. Other alterations focus on inducing sterility in the pest organism or affecting the digestion or metabolism of the pests.

In addition, attempts to enhance a plant's natural ability to produce leaf wax could make the plant more difficult for insects to consume. The best known and most widely used transgenic pest-protected crops are those that express insecticidal proteins derived from genes cloned from the soil bacterium Bacillus thuringiensis, more commonly known as Bt. Crystal (Cry) proteins or delta-endotoxins formed by this bacterium are toxic to many insect species.

Delta-endotoxins bind specifically in the insect gut to receptor proteins, destroying cells and killing the insect in several days. There are several different Bt strains containing many different toxins. Scientists have identified and isolated the genes for several toxin proteins from different Bt strains.

In recent years, these genes have been introduced into several crop plants in an effort to protect them from insect attack and eliminate the need for spraying synthetic chemical pesticides. There are more than 100 patents for Bt Cry genes. Bt field corn, sweet corn, soy, potato and cotton are commercialized in the U.S., and one or more of these are commercialized in at least 11 other countries.

Bt controls the larvae of butterflies and moths (Lepidopteran insects) that eat the plants. It is especially effective against the larvae of the European corn borer, a significant corn pest in the U.S., as well as the Southwestern corn borer and the lesser cornstalk borer. In sweet corn, Bt toxins effectively deter corn earworm and fall armyworm.

Recently, a different strain of Bt, Bacillus thuringiensis tenebrionis, was used as a gene source to confer resistance to corn rootworm, another major pest in cornfields. The resistant corn is currently in field trials. Bt hybrid rice is also undergoing field-testing and is showing considerable effectiveness in resisting major pests in Asia such as the leaf folder, yellow stem borer and striped stem borer.

Bt canola is also under development. Borers also create a good environment for fungi to grow. Where fusarium fungi grow, they reduce plant quality and generate fumonisins-toxins that can be fatal to farm animals and have been linked to liver and esophageal cancer in African farmers. Thus, one way to reduce fungal contamination is to control pests.

Scientists have measured reductions in fumonisin levels in Bt corn of 90 percent or greater. Bt works against insects that eat plant tissue. However, those pests that do not eat the leaves, but rather pierce and suck nutrients from the plant, require different defence strategies. These insects include aphids, white flies and stink bugs.

White flies are a major pest in poinsettias, sweet potatoes and cotton. Because these insects do not consume large amounts of plant material, a leading way to combat them is the genetic expression of toxic proteins that are strong enough to kill the pest, yet safe for the plant and non- target organisms.

Avidin in transgenic corn demonstrates a different approach. Avidin is a glycoprotein, an organic compound composed of both a protein and a carbohydrate, and is usually found in egg whites. Avidin is known for chemically tying up the vitamin biotin, making it unavailable as a nutrient. Insects eating transgenic corn modified to produce avidin die from biotin deficiency.

Although this corn was not toxic to mice, further evaluation of its potential for insect toxicity and safety for human consumption is awaited. Plants produce wax as a natural protective coating. Genetic modification can increase the expression of this inherent trait.

Experiments to increase leaf wax are in the early stages, but scientists have already raised wax content by as much as 15-fold. This strategy is aimed at increasing the plant's resistance to both pests and fungal pathogens.

Developing Nematode Resistance Food Crops

The most common of nematode plant parasites found worldwide is the root-knot nematode. Probably every form of plant life, including field crops, ornamentals and trees, is attacked by at least one species of nematode. They are responsible for 10 per cent of global crop losses worth an estimated $80 billion a year.

Transgenic strategies to combat nematodes are emerging. Nematodes are particularly destructive in bananas, soybeans, rice and potatoes. Scientists are fighting these parasitic worms in potato and banana crops using the genes for cystatins, defence proteins that occur naturally in rice and sunflowers. Incorporation of the genes in potatoes produced as much as 70 per cent nematode resistance in field trials.

Nematodes are particularly fond of soybeans. In the U.S., the soybean cyst nematode is considered the most devastating pest. Standard plant breeding led to a highly resistant variety of soybeans from a wild strain, but it did not cross well with modern soybean lines.

Using genetic markers, a means of identifying cells with particular traits, scientists bred plants containing the resistance gene with domesticated varieties, circumventing the poor performance characteristics of the wild variety. While the new varieties are not transgenic, they resulted from combining the use of modern genetic markers with conventional breeding techniques.

Improving Field-crop Production and Soil Management

Improving field-crop production and soil management is another central aim of genetic engineering technology in commodity crops. Applications include crop resistance to herbicides; improved nitrogen utilisation, reducing need for fertiliser; increased tolerance to stresses such as drought and frost; regulation of plant hormones, which are key to plant growth and development; attempts to increase yield, and a multitude of other, less widespread applications.

There are many negative effects when weeds grow with crop plants, the most common being competition for sunlight, water, space and soil nutrients. If weeds grow with crops, they too use these growth factors, and may cause losses great enough to justify control measures.

In addition to economic yield loss, other concerns may determine when weed control is justified. For example, eastern black nightshade in soybeans or late-emerging grasses in corn may not reduce yield, but these weeds can clog equipment, causing harvest delays. The most common method currently employed to manage weeds is the use of herbicides.

The use of genetic modification techniques has created crops that are both tolerant and resistant to herbicides, or weed killers. This technology allows herbicides to be sprayed over resistant crops from emergence through flowering, thus making the applications more effective.

To date, six categories of these crops have been engineered to be resistant to the herbicides glyphosate, glufosinate ammonium, imidazolinone, sulfonylurea, sethoxydim and bromoxynil.

Probably the best-known herbicide for which tolerance has been genetically engineered into crops is glyphosate, known commercially by brand names such as Roundup, Rodeo and Accord. Resistance to glyphosate is the transgenic trait most common in agriculture worldwide. To date soy, corn, cotton, canola, sugar beets and, most recently, wheat, have been genetically transformed for glyphosate tolerance.

Although glyphosate has been used as an herbicide for 26 years, transgenic glyphosate-resistant crops are a more recent development and are widely deployed on acres devoted to soy and cotton. Research is underway to create other glyphosate tolerant crops. To date, two weed species, annual rigid ryegrass and goose grass, have built resistance to glyphosate.

Corn, soy, rice, sugar beet, sweet corn and canola have also been genetically modified to tolerate the herbicide glufosinate ammonium. The seeds for these crops are sold commercially under brand names such as Liberty Link. Transgenic soybeans, cotton and flax with a tolerance to the herbicide sulfonylurea are also on the market.

Other strains of engineered soybeans and corn are resistant to sethoxydim, the active ingredient in the commercial herbicides Poast, Poast Plus, and Headline, used to control undesirable grass species.

The herbicide bromoxynil, sold under the commercial name Buctril, is normally toxic to cotton, a broadleaf crop, and is primarily used on grass-like crops, such as corn, sorghum and small grains, to kill invading broadleaf weeds. Scientists have genetically modified cotton plants for resistance to this herbicide, allowing its use to control broadleaf weeds in cotton fields.

Improved Nitrogen Utilization

There appear to be relatively few biotechnology applications specifically designed to enhance the characteristics of farm crops, such as size, yield, branching, seed size and number. Scientists have, however, created some enhancements. A recent example is the discovery of a gene in the alga Chlorella sorokiniana that has a unique enzyme not found in conventional crop plants.

The enzyme, ammonium-inducible glutamate dehydrogenase, increases the efficiency of ammonium incorporation into proteins. In some plants, it increases the efficiency of nitrogen use. The practical implication is that less fertilizer would be necessary to grow these plants. When the gene was incorporated into wheat, biomass production, growth rate and kernel weight all increased, as did the number of spikes in the plant.

Stress Tolerance

Stress tolerance involves a family of genes, rather than a single one. They are rapidly activated in response to cold, inducing the expression of "cold-regulated" genes, and resulting in enhanced freezing tolerance.

Over-expression of these genes in Arabidopsis-small plants of the mustard family that are commonly used to study plant genetics-increases freezing tolerance and leads to elevated levels of proline and total soluble sugars, substances that protect against cold. Common stress responses in plants involve water retention at the cellular level.

As a result, researchers have given special attention to osmoprotectant molecules, or molecules that hold water, such as sugars, sugar alcohols, certain amino acids (proline) and quaternary amines like glycinebetaine.

Various plants genetically engineered for increased levels of protectant sugar have shown increased drought tolerance. For instance, Arabidopsis and tobacco plants engineered to produce mannitol, a sugar alcohol, withstood high saline conditions and had enhanced germination rates and increased biomass. Other strategies have addressed different stress factors.

Improved cold tolerance and normal germination under high salt was reported in Arabidopsis engineered to express the enzyme choline oxidase. Transgenic rice, engineered to express the late embryogenesis abundant protein gene transferred from barley, was significantly more tolerant to drought and salinity than conventional varieties of rice.

Another transgenic rice engineered in the laboratory for enhanced expression of the enzyme glutamine synthetase had increased photorespiration capacity and increased tolerance to salt. Preliminary results suggested enhanced tolerance to chilling as well.

Regulation of Plant Hormones

Plant hormones such as auxin, cytokinins, gibberellins, abscisic acid, ethylene, etc. have been targeted for genetic modification to influence plant growth and development- fruit development and ripening; stem elongation and leaf development; germination, dormancy and tolerance of adverse conditions.

These hormone classes are highly interactive; the concentration of one affects the activity of another. For example, the ratio of the hormone abscisic acid to gibberellin in a plant determines whether a seed will remain dormant or germinate.

Recent discovery of an enzyme involved in the production of the hormone auxin enabled researchers to investigate the effects of moderating auxin production in determining plant characteristics. When auxin is overproduced, branching is inhibited and leaves curl down as the plant elongates, a reaction typically related to reduced light exposure. The same gene that produces this enzyme is apparently related to a gene in mammals that governs enzymes that detoxify certain chemicals.

In wheat, the hormone abscisic acid slows seed germination and improves the tolerance to cold and drought. Extending or enhancing the production of abscisic acid may also delay germination, a useful characteristic in climates where spring rain is sparse or falls late in the season.

Production of abscisic acid is increased in response to environmental stress, and a family of enzymes called protein kinases stimulates its production. Selecting plant varieties high in abscisic acid, or engineering plants to produce more of the hormone, may confer greater drought and cold tolerance.

Introduction of dwarfed, high-yielding wheat contributed to the 'Green Revolution' of the 1960s and 1970s, during which world wheat yields almost doubled. Shorter varieties of wheat grains, with a greater resistance to damage by wind, resulted from a reduced response to the hormone gibberellin.

Scientists have since shown that the gene called Rht can cause "dwarfing" in a range of plants, opening up the possibility of quickly developing higher-yielding varieties in several crops. Researchers believe that this strategy could be applied to a still wider range of crops through genetic engineering.

The plant hormone ethylene regulates ripening in fruits and vegetables. Controlling the amount and timing of ethylene production can initiate or delay ripening, which might reduce spoilage that can occur between the time produce is picked and brought to market.

Transgenic techniques aim to regulate the enzyme that breaks down a precursor of ethylene production. By regulating the timing and rate of this degradation, ripening can be controlled. This technology has been applied and field- tested in tomatoes, raspberries, melons, strawberries, cauliflower and broccoli, but has not yet been commercialized.

References

- What-is-agricultural-biotechnology, mixer-direct-blog: mixerdirect.com, Retrieved 28 April, 2019

- Plant-breeding, strategies-for-enhancement-in-food-production, biology, guides: toppr.com, Retrieved 21 April, 2019

- Methods-of-gene-transfer-in-plants-2-methods, engineering, genetics: biologydiscussion.com, Retrieved 17 February, 2019

- Transgenic-plants, meaning, reasons-and-fundamentals, transgenic-plants: biotechnologynotes.com, Retrieved 24 March, 2019

- Applications-of-transgenic, plants-6-applications, transgenic-plants, plants: biologydiscussion.com, Retrieved 16 June, 2019

- Disease-resistance-in-plants-5-techniques-biotechnology, genetically-modified, biotechnology: biotechnologynotes.com, Retrieved 13 July, 2019

Animal Biotechnology

The branch of biotechnology which makes use of molecular biology techniques to genetically engineer animals is known as animal biotechnology. Transgenic animals are a category of genetically engineered animals whose genome has been modified by the transfer of a gene from another species. The diverse applications of animal biotechnology have been thoroughly discussed in this chapter.

Animal biotechnology has been defined in various ways. One of the most recent quoted definitions of animal biotechnology is: "The application of scientific and engineering principles to the processing or production of materials by animals or aquatic species to provide goods and services".

Animal Biotechnology has developed rapidly since early 1980s when the first transgenic mice and first in vitro produced bovine embryos were created.

Examples of animal biotechnology include generation of transgenic animals or transgenic fish using gene knockout technology to generate animals in which a specific gene has been inactivated, production of nearly identical animals by somatic cell nuclear transfer, or production of infertile aquatic species.

To cater to the need of our fast growing population, searching cheap sources of proteins either from animals or from plants have been the most daunting task for us. Animal protein is an excellent source of dietary amino acids essential for human nutrition. In addition, we are able to harvest meat, milk and fibre from live animals and use them for different welfare programmes. The production of food from animals is expensive and less efficient than plants. In spite of these disadvantages, we continue to farm animals on commercial basis for food, fibres and by-products.

Our ancestors first domesticated animals many thousands of years ago and since then they have been our constant companions. From that time onwards, the animals themselves started to change. With the passage of time and continued isolation from their wild relatives, domestic animals started to display different characteristics. Some would have produced more or better milk than others, some would have grown faster or been more fecund, some would have been more vigorous.

Our ancestors might .have noticed these variations and mated animals with similar desirable traits, but they did so without understanding the basis of heredity. Restrictive breeding and selective mating gradually led to the many diverse breeds of livestock we have today, although most breeds of the dominant domestic cattle, Bos taurus, arose in the last couple of centuries. Herds of ancestral species, Bos primigenius, lived in Europe in the 18th century.

Thus animal improvement is no longer a matter of interest but it is a matter of necessity. Biotechnology provides us powerful tools that engage us learning the basics of these tools for increase in animal production world wide.

Reproduction in Animals/Reproductive Technology

Reproduction has some important roles in production system. It must occur at a rate that ensures replenishment of stock is greater than its use and improving genetic quality of stock. The aim of reproductive technologies is to increase the number of progeny while improving the genetic quality of stock in general and most intensively managed animals like dairy cattle and domestic pigs in particular. This has been done by various procedures like artificial insemination, embryo transfer, in vitro fertilization and embryo cloning.

Artificial Insemination (AI)

Artificial insemination, the first animal biotechnology, takes advantages of the male's excess gametes (sperm) production capacity. This has been the single-most important factor in increasing the productivity. There are many variations of artificial insemination to consider. There is animal-animal artificial insemination, fresh-extended (chilled) semen insemination and frozen semen insemination.

Before insemination the sperm ejaculated is collected, diluted and examined under microscope to determine the number and mortality of sperms. Techniques of AI include vaginal deposition, surgical implant, and trans-cervical insemination. Studies have been published with convincing evidence that every semen preparation in fresh, fresh-extended or frozen will produce larger litters if semen is deposited into the uterus, especially with trans-cervical technique. Fresh and fresh-extended semen produce good results with vaginal deposition.

For cattle, AI is done in standing animals without using anesthesia, using a procedure known as "rectal palpation". Of the different types of semen preparation, it is again obvious that fresh animal-animal collection and insemination will give the best semen preparation. Fresh-extended or "chilled" semen has performed very well for many years, but not that well as does the fresh semen.

The worst results are with the frozen semen, no matter what process is used for the freezing. Females are inseminated after ovulation, which last only a few hours and occurs most commonly at night. In many breeds of farm animals it is very difficult to detect ovulation (oestrous).

The alternative is to induce females to ovulate in a synchronized manner. In practice, it is rather impossible to achieve total synchrony of ovulation but about 80% of females would respond to the inducer. With ruminants, ovulation in females can be induced by administration of hormone like progesterone or prostaglandin.

Single sex is being preferred by all livestock industries. In dairy and beef cattle, females are more accepted than males because females are not efficient in reproduction and having desirable production traits. Same thing happens in pig industry. The sex of the embryo is determined solely by sperm containing either X chromosome or Y chromosome.

Since half of the sperm in an ejaculate contain the X and the remaining half the Y, the sex ratio of progeny is near to 1 : 1. Separating sperm containing the X chromosome from that of the Y would allow us to manipulate the sex progeny by AI. Separation of the two populations has been done using a fluorescent activated cell sorter (FACS). The method is very expensive and very slow too. Monoclonal antibody that binds specifically to Y-bearing sperm cells could be used in future to separate Y- bearing and X-bearing sperm. The flow sorting of sperm cells to separate X from Y bearing cells has been successful in most cases tested and in cattle and swine has resulted in off-spring of the desired sex.

Embryo Transfer

The propagation of genetically valuable animals is enhanced by transferring embryo. To do this one has to increase the number of mature eggs produced by a selected female. Following fertilization, those fertilized eggs are implanted into foster mother, one for each egg. Donor females are injected with prostaglandin F2a (PGF2a) to induce a synchronized oestrous before treatment is started. Ten days after oestrous period they are injected with FSH for a period of four days, followed by PGF2a to induce oestrous.

Superovulated donors are then mated by artificial insemination. Six to eight days after insemination, the fertilized eggs are recovered from cattle and six days from sheep and goat. They are immediately transferred into synchronized recipients. Alternatively, they can be frozen for indefinite storage. With good practice, pregnancy rates of 50-60% can be achieved in cattle from transferred embryo.

Pregnancy output rates can be increased by embryo splitting; a method which produces identical twins. Recently this has become a routine production tool, requiring micromanipulation equipment and minimal training. Embryos were obtained from donor at the blastocyst stage and transferred into a standard cell culture medium containing hypertonic sucrose and bovine serum albumin.

The embryo is then transferred to a plastic Petridish containing standard culture medium, where it sinks to the bottom and sticks to it. It occurs because of the electrostatic interactions between the BSA-induced negative charge on the outer membrane and the positive charge of the plastic disc.

The embryo is then bisected using a micromanipulator fitted with a fine blade under the inverted microscope. During embryo splitting care must be taken to ensure that the inner cell mass is split into equal halves. After splitting, they are transferred into the oviduct of the recipient female following the same procedure as for normal transfer. Routine biopsy has been carried out to determine the sex of progeny, a production trait. Presence of Y chromosomal DNA in cells of an embryo biopsy indicates that the embryo is a male: if no Y chromosomal DNA is present it is female. This is being done using PCR to amplify Y-chromosome specific DNA.

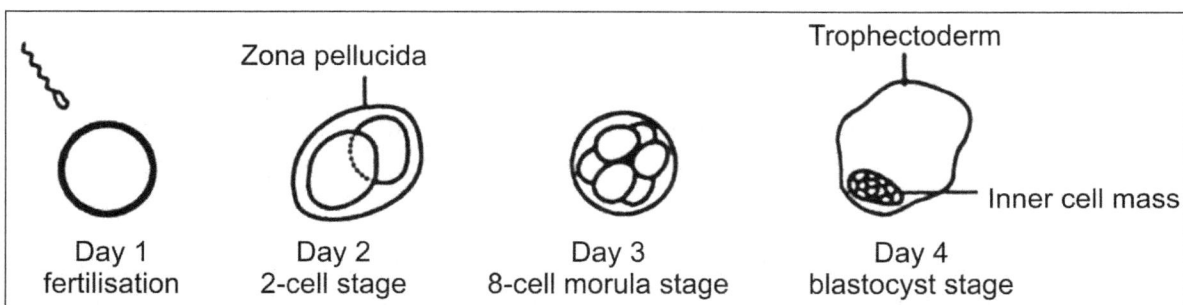

Embryo Development at the Early Stages.

In Vitro Fertilization (IVF)

Embryo transfer is not used widely because of many factors including high costs, technical difficulty and limited availability. We can overcome these problems if oocytes, collected from a donor female are fertilized in vitro. It would ensure fertilization of more eggs using less amount of semen. Thus IVF is more efficient method than AI. During oestrous cycle a number of Graafian follicles start growing.

Eventually one of the follicle matures and ruptures, releasing the eggs for fertilization. Oocytes are recovered from these follicles from super ovulated donors by laparoscopic surgery. Eggs are then incubated to be matured and fertilized in vitro. This method demands high technical skill and, therefore, not always cost effective.

The success of IVF depends on collection of large number of oocytes and there maturation in vitro. This can be done by removing an ovary from a selected female and is induced to mature in vitro. Numbers of scientists across the globe have been carrying out research on oocyte maturation in vitro. However, it may be worthwhile to mention here that majority of Graafian follicles never attain maturity. The highest success rates with embryo transfer are reported when blastocysts are implanted into recipients.

This can be done if embryos are maintained in culture. Many laboratories have shown that 60% of IVF cattle embryos can be cultured to blastocyst stage. High rate of abortion is recorded from culture embryos during the first two months of pregnancy. This may be due to genetic defect in the oocyte and or fertilizing sperm and environmental mutagenesis of egg, sperm or embryo (mainly caused by oxygen free radicals).

Cloning

Another application of animal biotechnology is the use of somatic nuclear transfer to produce multiple copies of animals that are merely identical copies of other animals. This process is called 'cloning'. So cloning is the production of one or more identical plants or animals that are genetically identical to another plant or animal.

It would facilitate increased production of numbers of one particular embryo having desirable characteristics. Nature itself is the greatest cloning agent. In about one of every 75 conception, the fertilized embryo splits and produces monozygotic twins.

DNA Cloning

To do further manipulations and analysis of individual recombinant DNA molecules (e.g. DNA containing insulin gene), we must have many copies of the molecules, usually in a purified form. DNA cloning is a technique to produce large quantities of specific DNA segment. The DNA segment to be cloned is inserted to a vector, which is a vehicle for carrying inserted DNA into a suitable host cell, such as the bacterium E. coli.

The vector contains sequences that allow it to be replicated within the host cell and usually refer to as cloning vector. There are numerous cloning vectors in current use and choice among them depends on the size of the DNA fragment that needs to be cloned or the use to which the clone will be put.

Plasmid Vectors

Bacterial plasmids are small circular DNA molecules that are distinct from the main bacterial chromosome. They replicate their DNA independently of the bacterial chromosome. The plasmids that are routinely used as vectors are those that carry genes for drug resistance. The drug resistance genes are useful because drug- resistant phenotype can be used to select for those cells that contains the recombinant plasmid.

The process by which bacterial cells take up DNA from the medium is called transformation. This forms the basis for cloning plasmid in bacterial cells. In most commonly used methods recombinant plasmids are added to a bacterial culture that has been pretreated with Calcium ions. Bacterial cells are then activated (by heat-shock) to take up DNA from the medium.

Normally a small number of cells are able to take up and retain one of the recombinant plasmid molecules. Once a bacterial cell has taken up a recombinant plasmid from the medium, the cell gives rise to a colony of cells containing the recombinant DNA molecule. These bacteria containing a recombinant plasmid are selected from the rest by growing the cells in the presence of the antibiotic specific to drug-resistance gene of plasmids.

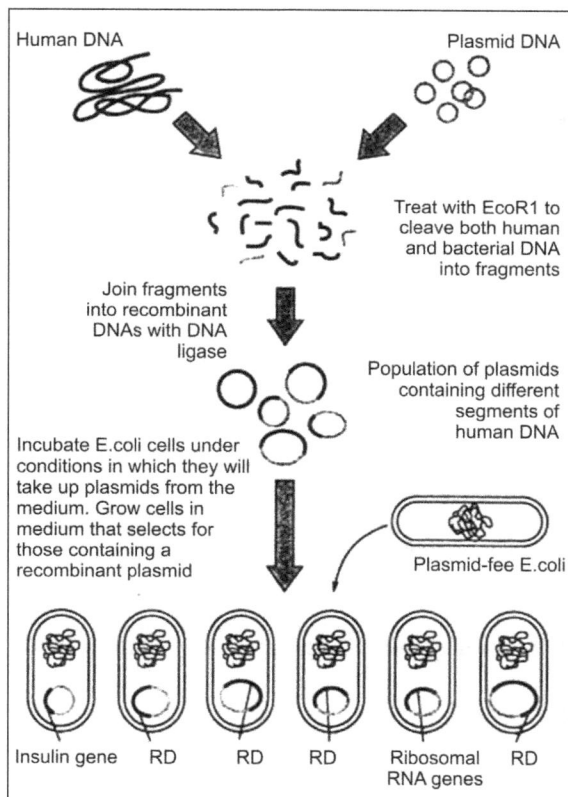

DNA Cloning Using Plasmid Vectors.

Bacteriophage Vectors

There are different classes of bacteriophage vectors, depending on whether chromosomal DNA inside the bacteriophage is single stranded or double stranded and the size of the donor DNA insert.

Bacteriophage (Lambda)

It is used as a cloning vector for double-stranded DNA inserts up to approximately 25 kb. Lambda phage heads harbor a linear DNA (genome) of about 50kb in length. The central part of the genome is not required for replication or packaging and so a central part can be cut out by using restriction enzyme and discarded.

The two remaining 'arms' at either end of the genome are ligated to restriction-digested donor DNA. The recombinant molecules can be introduced into E. coli by transformation. Once in the bacteria, the donor DNA insert is amplified along with the lambda/phage DNA and packaged into a new generation of virus particles, which are released when the cell is lysed. The released particles infect new cells. This results into the occurrence of a clear spot (or plaque) in the bacterial plate at the site of infection. Each plaque harbors millions of large particles, each carrying a single copy of same donor DNA insert.

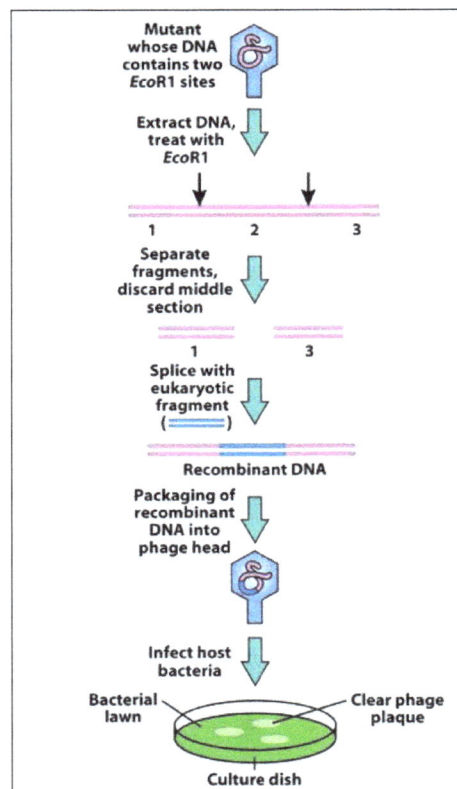

Cloning Eukaryotic DNA Fragments in Lambda Phage.

Vectors for Larger DNA Inserts

The maximum size of the donor DNA that can be inserted into a standard plasmid or vector is about 30 kb in length. To meet these demand, several vectors have been engineered. The largest prokaryotic inserts use the BAC (bacterial artificial chromosome) vector system. It is based on the 7-kb F plasmid and has the ability to accept larger DNA inserts (up to about 300 kb).

For inserts larger than 300 kb, YAC (yeast artificial chromosome) — a eukaryotic vector system based on yeast chromosomes introduced into yeast cells by transformation, is being used to clone recombinant molecules as large as 1,000 kb in length.

Formation of a DNA Library

DNA cloning is frequently used to produce DNA libraries, which are collections of cloned DNA fragments. There exists two basic types of DNA libraries: genomic libraries and cDNA libraries. Genomic libraries are produced from total DNA obtained from nuclei and contain all of the DNA sequences of the species.

Once a genomic library of a species is created, scientists can use the library to isolate specific DNA segments. cDNA libraries are derived from DNA copies of mRNA population. cDNA libraries are produced from mRNA present in a particular cell type and this corresponds to the genes that are active in that type of cell.

Identification of DNA Molecules of Interest

Immediate after cloning, the next task is to find that particular clone which contains the desired gene. This has been done either by using a specific probes (for finding DNA or protein) or by probing a specific nucleic acid.

Use of Recombinant DNA Technology for Genetic Engineering

The use of sophisticated recombinant DNA techniques to alter the genotype and phenotype of an organism is called genetic engineering. This is being done by introducing genes into eukaryotic cells, where they are transcribed and translated. There are number of strategies to achieve this. The most frequently used technique is the viral-mediated gene transfer, which is termed transduction.

Here the engineered DNA is incorporated into the genome of a non-replicating virus and allow the virus to infect the cell. It is the type of virus that would determine whether gene of interest can be expressed temporarily or integrated into the genome of the host cells. Retroviruses have been used in gene therapy to transfer a normal gene into the cells of a patient having a defective gene. Transfection is a process by which naked DNA can be introduced into cultured cells.

During the process the cells are treated with either calcium phosphate or DEAE-dextran, both of which form a complex with the added DNA that promotes its adherence to the cell. It is observed that only a few cells take up the DNA and incorporates it stably into the chromosomes to make a transgenic eukaryote. Electroporation and lipofection—the two other methods have also been used to transfect cells.

In lipofection, foreign DNA binds to positively charged lipids (liposomes) which are capable of fusing with lipid bilayer of the cell membrane and supplying the DNA to the cytoplasm. In electroporation, foreign DNA finds their way through the modified plasma membrane, induced by electric current, into the nucleus and become integrated into the genome. Foreign DNA can also be introduced into a cell by microinjecting directly into the cell nucleus.

For a long time Xenopus oocytes have been used to study the expression of foreign genes. It contains all ingredients for mRNA synthesis. When foreign DNA is delivered into the nucleus, it initiates transcription and eventually m-RNA is transported to the cytoplasm, where they are translated into proteins that can be detected immunologically.

Another important target for injected DNA is the nucleus of a mouse embryo. Here the foreign DNA become integrated into the egg's chromosome, which will pass on to all the cells of the embryo and finally to the adult. Those animals that have been genetically engineered are called transgenic animals.

The first transgenic animal was created by Ralph Brinster of the University of Pennsylvania and Richar Palmiter of the University of Washington in the year 1981. They succeeded in introducing a gene for rat growth hormone (GH) into the fertilized eggs of mice. The injected DNA was constructed in such a way that the rat GH gene (coding protein) is located downstream from the promoter region of the mouse metallothionein gene.

In a normal rat, the synthesis of metallothionein gene is accelerated following the treatments of metals like cadmium, zinc, or glucocorticoid hormones. In the transgenic mice, synthesis of the GH gene has seen to have enhanced following treatment with metals and glucocorticoids. Mice are the most important models for mammalian genetics. The technology developed in mice is applicable to human.

There are two strategies for transgenesis in mice: ectopic insertion and gene targeting. The procedure involved in ectopic insertion is simply to inject bacterially cloned DNA solutions into the nucleus of early-stage embryos,. Gene targeting is a bit rare event and a multi-step process is needed involving the use of embryonic stem cells. Embryonic stem cells (ES) have the ability to form any and all parts of a mouse, that is why they are called totipotent cells.

The process of gene targeting can be recognized by one of its outputs, the substitution of a non-functional gene for the normal gene. This type of targeted inactivation is called a gene knockout. Here embryonic stem cells are isolated from a mouse strain (albino) and embryos are transfected with a DNA insert containing an inactive, mutant allele of the gene to be knocked out.

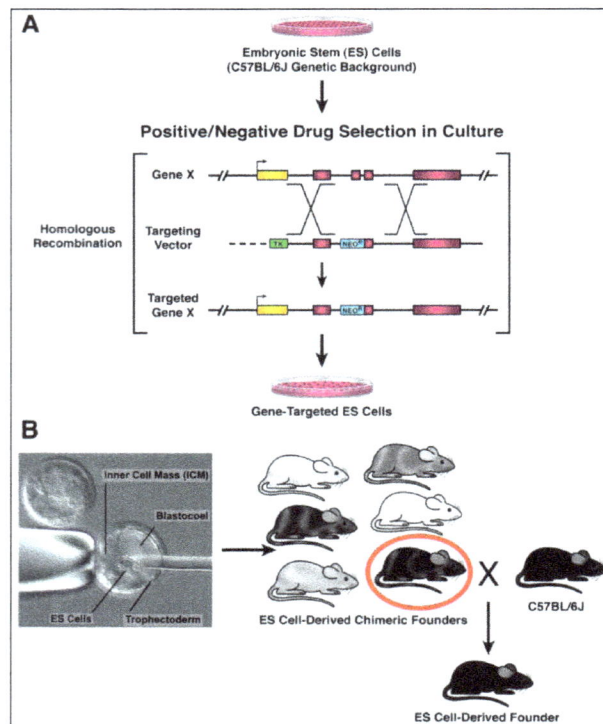

Formation of Knockout Mice.

The ES cells are then injected into the recipient young mouse embryo (blastocoel) collected from an albino strain. In the next step, the embryo containing the ES cells grows to term in surrogate mother. The resulting progeny are chimeric, having tissue derived from both the donor (transplanted ES) and recipient strains. Chimeric mice are then mated with their sibling to produce homozygous mice with the knock-out in each copy of the gene. These are knock-out mice that lack a functional copy of the gene.

The term 'cloning' usually refer to three different procedures. The three types of cloning are: embryo cloning, adult DNA cloning, and therapeutic cloning.

Embryo Cloning

It might be more accurately called 'artificial twining' because it stimulates the mechanism by which twins naturally develop. It involves removing one or more cells from an embryo and encouraging a cell to develop into a separate embryo with the same DNA as the original.

It has been successfully carried out for years on many species of animals including cattle, sheep, pigs, goats, horses, moles, cats, rats and mice. The technology of embryo cloning has been done in two ways: nuclear transfer and embryonic stem cell. Cloning of embryos has been used in mice since the late 1970s and animal breeding since 1980s.

Nuclear Transfer

The technique involves culturing somatic cells from an appropriate tissue (preferably fibroblast) from the animals to be cloned. Nuclei from the cultured somatic cells are then micro-injected into an enucleated oocyte obtained from another individual in same or closely related species.

Through a process that is not yet understood, the nucleus from the somatic cells is reprogrammed to a pattern of gene expression suitable for directing normal development of the embryo. After further culture and development in vitro, the embryos are transferred to a recipient female, and, ultimately, would result in the birth of live offspring.

The success rate of producing animals by nuclear transfer is less than 10% and depends on many factors including the species involve, source of recipient ova, cell type of donor nuclei, treatment of donor cells prior to nuclei transfer and techniques used for nuclear transfer.

The first publicly announced human cloning was done by Robert J. Stillman and his team at the George Washington Medical Centre in Washington D.C. They took 17 genetically defective human embryos. These embryos were derived from ovum that had been fertilized by two sperms. This has resulted into an extra set of chromosomes which ruined the ovum's fate. None could have developed into a fetus. These ovum were successfully split in 1994, each producing one or more clones.

Adult DNA Cloning (Reproductive Cloning)

This technique is used to produce a duplicate of an existing animal. It has been used to clone a sheep and other mammals. Reproductive cloning was earlier thought to be impossible in all mammals until it was activated in 1996 by a scientist, Dr. Ian Wialmut of the Roslin Institute in Roslin, Scotland, U. K. "Dolly", a seven-month old sheep, was displayed to the media on February 23rd, 1997. She is the first large cloned animal using DNA from another adult.

A cell was taken from the mammary tissue of a mature six-year old sheep while its DNA was in dormant state. It was fused with sheep ovum which had its nucleus removed. The 'fertilized' cell was then stimulated with an electrical pulse. Out of 277 attempts at cell fusion, only 29 began to divide. These were all implanted in ewes, 13 became pregnant but only one lamb, Dolly, was borne. On March 4, 1997, President Clinton ordered a widespread ban on the Federal funding on human cloning in the USA. Research on human cloning continues in other countries.

Over the past 25 years tremendous advances have been made in the analysis of eukaryotic genomes. This progress began as molecular biologists learned to construct recombinant DNA molecules, which are molecules containing DNA sequences derived from more than one source.

It involves the isolation from the genome of a particular segment of DNA that codes for a particular polypeptide, isolation of enzymes which would cut DNA at precisely defined locations (restriction endonucleases) and those which would covalently join DNA fragments (ligases).

The main steps of DNA cloning are given and involve the following procedures:

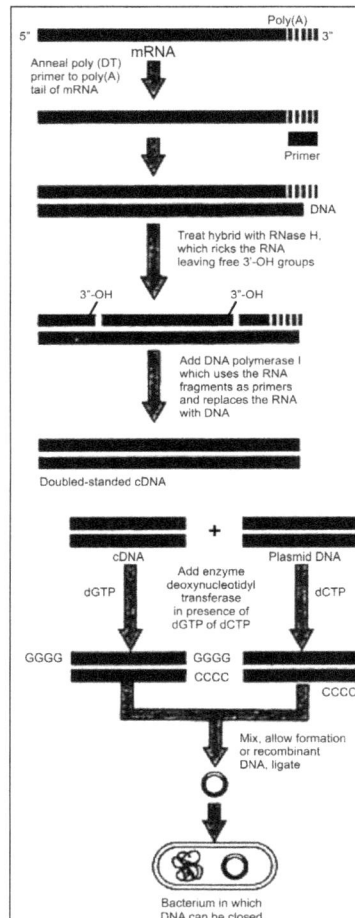

Synthesis of cDNAs for Cloning.

1. Isolation of the DNA to be cloned.

2. Insertion of the isolated DNA into another piece Of DNA called a vector, which is a vehicle for carrying foreign DNA into suitable host cell such as the bacterium E. coli. Although.

shows the use of a plasmid vector, other types of vector, can also be used such as the Bacteriophage lambda.

3. Transfer of the recombinant vectors in bacterial cells either by transformations by infection using viruses.

4. Selection of those cells which contain the desired recombinant vectors.

5. Growth of the bacteria to give as much cloned DNA as is needed.

Isolation of DNA to be Cloned

The organism under study, which will be used to donate DNA for the analysis, is called the donor organism. Three sources of DNA segment can be considered: genomic DNA, complementary DNA and chemically synthesized DNA.

Genomic DNA

These DNA sequences exist in the chromosomes of the organism under study and thus are the most easy source of DNA.

Complementary DNA (cDNA)

Complementary DNA or cDNA, is essentially a double-stranded DNA version of an mRNA molecule. cDNA is made from mRNA with the use of a special enzyme called reverse transcriptase, originally isolated from retroviruses. With the use of an mRNA molecule as a template, reverse transcriptase synthesizes a single-stranded DNA molecule that can be used as a template for double- stranded DNA synthesis. Synthesis of cDNA is very important in the analysis of gene structure and gene expression.

Chemically Synthesized DNA

Techniques for the chemical synthesis of oligonucleotides have been developed to produce DNA sequence for the purpose of recombinant DNA.

Construction of Recombinant DNA

Isolated DNA molecules are broken into small fragments by enzymatically digesting it with endonucleases (restriction enzymes) that cleave at specific sites. Restriction enzymes cut at specific DNA target sequences (involving 4 to 8 nucleotides) and this property is one of the key features that make them suitable for DNA manipulation.

It is purely, by chance, any DNA molecule, be it derived from viruses, bacteria, plant and animal, contains restriction enzyme target sites. Thus, in the presence of appropriate restriction enzyme, the DNA would be cut into a set of small fragments according to the location of the restriction sites.

The Restriction Enzyme EcoR1 (from E. coli) Recognizes the following 6-Nucleotide-Pair Sequence in the DNA of Any Organism:

1. 5'-GAATTC-3'

2. 3'-CTTAAG-5'

This type of segment is called a DNA palindrome, which means that both strands have the same nucleotide sequence but in antiparallel orientation. EcoR1 recognizes and cuts only within the GAATTC palindrome sequence.

The cuts are between the g and the a nucleotides on each strand of the palindrome:

- 5'-G-AATTC-3'

- 3'-CTTAAG-5'

This staggered cuts leave a pair of identical five-base long single stranded 'sticky ends'. The ends are called sticky because, being single stranded, they can be hybridized through base-pair hydrogen bonding to a complementary copy. The production of this sticky ends is another feature of many restriction enzymes that makes them suitable tools for recombinant DNA technology.

If two DNA molecules are cut with the same sticky end-producing restriction enzyme, the fragments of each will have the same complementary sticky ends, enabling them to hybridize with each other under the appropriate conditions in a test tube. illustrates the restriction enzyme making a single cut in the circular DNA molecule such as a plasmid (vector); the cut opens up the circle, and the resulting a liner molecule has two sticky ends.

If such a molecule is mixed with a different DNA molecule (donor DNA fragment containing insulin—gene cut with EcoR1, as shown, the two can then hybridize to each other through complementary sticky- ends to form a recombinant molecule. There are two reasons why the donor DNA containing insulin gene must be attached to plasmid DNA segments to form useful recombinant molecules.

First, a fragmented donor DNA segment does not have a necessary DNA sequences, on its own, to enable it to be replicated in a test tube or inside a host organism. The donor DNA must be physically attached to other DNA segments that can support replication in a test tube or inside a host cell. Second, an experiment may demand that multiple fragments be glued together to form a functional unit (e.g. a transcriptionally active gene).

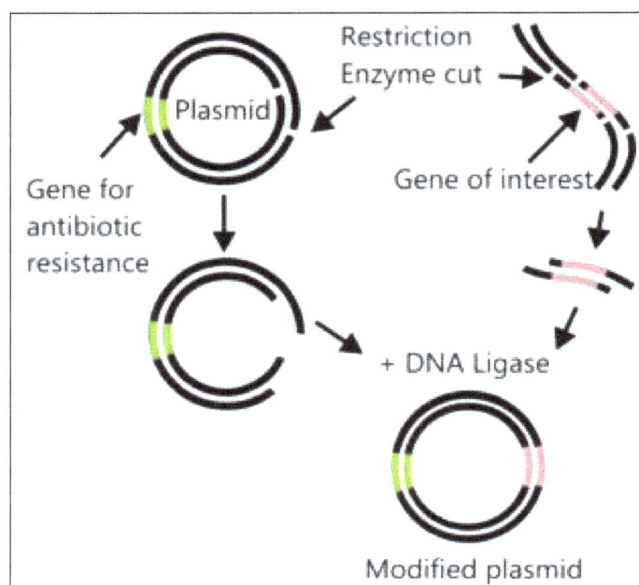

Formation of Recombinant DNA Molecule.

Most commonly, both donor and plasmid DNA are digested together in the presence of DNA ligase. During the incubation, two types of DNAs become hydrogen bonded to one another by their sticky ends and then ligated to form circular DNA recombinants.

Therapeutic Cloning

This is a procedure whose initial stages are identical to adult DNA cloning. However, the stem cells are removed from the pre-embryo with the interest of producing tissue or a whole organ for transplant back into the person who supplied the DNA. The pre-embryo dies in the process. The goal of therapeutic cloning is to produce a health copy of a sick person's tissue or organ for transplant.

The tissue or organ, would have the sick persons original DNA and, therefore, the patient need not have to take immunosuppressive drugs for the rest of life. Scientists are attempting to create transgenic pigs which have human genes. Their heart, liver or kidneys might be used as organ transplants in humans. This will save many lives because thousand of people die each year waiting for human organs.

Once achieved, transgenic animals could be cloned to produce as many organs as are needed. Researchers have produced transgenic animals typically in order to produce human hormones or proteins in its milk. These substances can be separated from the milk and be used for treating human ailments.

Transgenic Animals

There are various definitions for the term transgenic animal. The Federation of European Laboratory Animal Associations defines the term as an animal in which there has been a deliberate modification of its genome, the genetic makeup of an organism responsible for inherited characteristics.

The nucleus of all cells in every living organism contains genes made up of DNA. These genes store information that regulates how our bodies form and function. Genes can be altered artificially, so that some characteristics of an animal are changed. For example, an embryo can have an extra, functioning gene from another source artificially introduced into it, or a gene introduced which can knock out the functioning of another particular gene in the embryo. Animals that have their DNA manipulated in this way are knows as transgenic animals.

The majority of transgenic animals produced so far are mice, the animal that pioneered the technology. The first successful transgenic animal was a mouse.[6] A few years later, it was followed by rabbits, pigs, sheep, and cattle.

Why are these animals being produced? The two most common reasons are:

- Some transgenic animals are produced for specific economic traits. For example, transgenic cattle were created to produce milk containing particular human proteins, which may help in the treatment of human emphysema.

- Other transgenic animals are produced as disease models (animals genetically manipulated to exhibit disease symptoms so that effective treatment can be studied).

Importance of Transgenic Animals

The dependence of man on animals such as cattle, sheep, poultry, pig and fish for various purposes (milk, meat, eggs, wool etc.) is well known.

Trans-genesis has now become a powerful tool for studying the gene expression and developmental processes in higher organisms, besides the improvement in their genetic characteristics. Transgenic animals serve as good models for understanding the human diseases.

Further, several proteins produced by transgenic animals are important for medical and pharmaceutical applications. Thus, the transgenic farm animals are a part of the lucrative world-wide biotechnology industry, with great benefits to mankind. Trans-genesis is important for improving the quality and quantity of milk, meat, eggs and wool production, besides creating drug resistant animals.

Milk as the Medium of Protein Production

Milk is the secretion of mammary glands that can be collected frequently without causing any harm to the animal. Thus, milk from the transgenic animals can serve as a good and authenticated source of human proteins for a wide range of applications. Another advantage with milk is that it contains only a few proteins (casein, lactalbumin, immunoglobulin etc.) in the native state, therefore isolation and purification of a new protein from milk is easy.

Commonly used Animals for Transgenesis

The first animals used for trans-genesis was a mouse. The 'Super Mouse', was created by inserting a rat gene for growth hormone into the mouse genome. The offspring was much larger than the parents. Super Mouse attracted a lot of public attention, since it was a product of genetic manipulation rather than the normal route of sexual reproduction. Mouse continues to be an animal of choice for most transgenic experiments. The other animals used for trans-genesis include rat, rabbit, pig, cow, goat, sheep and fish.

Position Effects

Position effect is the phenomenon of different levels of gene expression that is observed after insertion of a new gene at different position in the eukaryotic genome. This is commonly observed in transgenic animals as well as plants. These transgenic organisms show variable levels and patterns of transgene expression. In a majority of cases, position effects are dependent on the site of transgene integration. In general, the defective expression is due to the insertion of transgene into a region of highly packed chromatin. The transgene will be more active if inserted into an area of open chromatin.

The positional effects are overcome by a group of DNA sequences called insulators. The sequences referred to as specialized chromatin structure (SCS) are known to perform the functions of insulators. It has been demonstrated that the expression of the gene is appropriate if the transgene is flanked by insulators.

Animal Bioreactors

Trans-genesis is wonderfully utilized for production proteins of pharmaceutical and medical use. In fact, any protein synthesized in the human body can be made in the transgenic animals, provided that the genes are correctly programmed. The advantage with transgenic animals is to produce scarce human proteins in huge quantities. Thus, the animals serving as factories for production of biologically important products are referred to as animal bioreactors or sometimes pharm animals. Frankly speaking, transgenic animals as bioreactors can be commercially exploited for the benefit of mankind.

Once developed, animal bioreactors are cost- effective for the production of large quantities of human proteins. Routine breeding and healthful living conditions are enough to maintain transgenic animals. A list of the therapeutically important proteins produced by animal bioreactors is given.

Transgenic Animals in Xenotransplantation

Organ transplantation (kidney, liver, heart etc.) in humans has now become one of the advanced surgical practices to replace the defective, nonfunctional or severally damaged organs. The major limitation of transplantation is the shortage of organ donors. This often results in long waiting times and many unnecessary deaths of organ failure patients.

Xenotransplantation refers to the replacement of failed human organs by the functional animal organs. The major limitation of xenotransplantation is the phenomenon of hyper acute organ rejection due to host immune system.

The organ rejections is mainly due to the following two causes:

i.   The antibodies raised against the foreign organ.

ii. Activation of host's complement system.

Pigs in Xenotransplantation

Some workers are actively conducting research to utilize organs of pigs in xenotransplantation. It is now identified that the major reason for rejection of pig organs by primates is due to the presence of a special group of disaccharides (Gal-α 1, 3-Gal) in pigs, and not in primates.

The enzyme responsible for the synthesis of specific disaccharides in pigs has been identified. It is α 1, 3-galactosyltransferase, present in pigs and not in primates. Scientists are optimistic that knockout pigs lacking the gene encoding the enzyme α 1, 3-galactosyltransferase can be developed in the next few years. Another approach is to introduce genes in primates that can degrade or modify Gal-α 1, 3-Gal disaccharide groups (of pigs). This will reduce immunogenicity.

Besides the above, there are other strategies to avoid hyperactive organ rejection by the hosts in xenotransplantation.

i.   Expression of antibodies against the pig disaccharides.

ii. Expression of complement— inactivating protein on the cell surfaces.

By the above approaches, it may be possible to overcome immediate hyperactive rejection of organs. The next problem is the delayed rejection which involves the macrophages and natural killer cells of the host.

Another concern of xenotransplantation is that the endogenous pig retroviruses could get activated after organ transplantation. This may lead to new genetic changes with unknown consequences.

The use of transgenic animals in xenotransplantation is only at the laboratory experimental stages, involving animals.

Methods of Creation of Transgenic Animals

For practical reasons, i.e., their small size and low cost of housing in comparison to that for larger vertebrates, their short generation time, and their fairly well defined genetics, mice have become the main species used in the field of transgenics.

The three principal methods used for the creation of transgenic animals are DNA microinjection, embryonic stem cell-mediated gene transfer and retrovirus-mediated gene transfer.

DNA Microinjection

This method involves the direct microinjection of a chosen gene construct (a single gene or a combination of genes) from another member of the same species or from a different species, into the pronucleus of a fertilized ovum. It is one of the first methods that proved to be effective in mammals. The introduced DNA may lead to the over- or under-expression of certain genes or to the expression of genes entirely new to the animal species. The insertion of DNA is, however, a random process, and there is a high probability that the introduced gene will not insert itself into a site on the host DNA that will permit its expression. The manipulated fertilized ovum is transferred into the oviduct of a recipient female, or foster mother that has been induced to act as a recipient by mating with a vasectomized male.

A major advantage of this method is its applicability to a wide variety of species.

The only successful method to produce transgenic rabbits, pigs, sheep, goats and cattle is DNA-microinjection technique into the pro-nucleus.

Rabbits

Rabbits are used as experimental models in gene transfer experiments. In 1985, the successful production of transgenic rabbits was reported, for the first time, and included the growth hormone construct MT-hGH. The rate of degeneration of rabbit zygotes caused by injection was below 10%. The pre-implantation development capacity of injected zygotes is significantly lower compared with control embryos.

Pigs

Pig zygotes must be centrifuged to show the pro-nucleus. Fifty percent of the centrifuged non-injected zygotes develop in vivo up to the morula or blastocyst stage. After microinjection, 10- 20% development to various stages of embryonic development occurs. Of the injected zygotes, 5.6% to 11% developed and led to the birth of piglets. The integration rate in pigs is approximately 10%. Growth-hormone constructs used in initial experiments led to an expression rate of 50%.

The production of transgenic F1 offspring is possible. In the authors' own experiments, inheritance of the trans-gene could be proved in two out of five animals. It is not necessary to centrifuge sheep embryos to make the pro-nucleus visible. According to "Nomarski optics", 80% of the pro-nuclei can be located if a microscope with interference contrast is available. The capacity for in vivo development of sheep zygotes with injection (26 per cent) and without (10 per cent) is half that of pig embryos after similar treatment.

Seven days after in vivo culture of non-treated and non-in vitro cultured sheep zygotes, Rexroad and Wall in 1987 observed a development rate of 86%. An in vitro culture of five hours' duration reduced this development rate to 65%, and after the injection of a buffer solution a reduction to 42 per cent was observed. 19% developed to the 32-cell stage after injection of DNA solution.

Sheep and Goats

Until recently, the trans-genes introduced into sheep inserted randomly in the genome and often worked poorly. However, in July 2000, success at inserting a trans-gene into a specific gene locus was reported. The gene was the human gene for alpha 1-antitrypsin, and two of the animals expressed large quantities of the human protein in their milk.

(1) It was done as sheep fibroblasts (connective tissue cells) growing in tissue culture were treated with a vector that contained these segments of DNA.

(2) Regions homologous to the sheep COL1 A1 gene. This gene encodes Type 1 collagen (Its absence in humans causes the inherited disease osteogenesis imperfecta). This locus was chosen because fibroblasts secrete large amounts of collagen and thus one would expect the gene to be easily accessible in the chromatin.

A neomycin-resistance gene to aid in isolating those cells that successfully incorporated the vector. The human gene encoding alphal-antitrypsin. Some people inherit two non- or poorly-functioning genes for this protein. Its resulting low level or absence produces the disease Alphal-Antitrypsin Deficiency (A1AD or Alphal).

The main symptoms are damage to the lungs (and sometimes to the liver).

(1) Promoter sites from the beta-Iactoglobulin gene. These promote hormone-driven gene expression in milk-producing cells.

(2) Binding sites for ribosome's for efficient translation of the mRNAs. Successfully-transformed cells were then:

- Fused with enucleated sheep eggs,

- Implanted in the uterus of a ewe (female sheep),

- Several embryos survived until their birth, and two young lambs have now lived over a year,

- When treated with hormones, these two lambs secreted milk containing large amounts of alphal-antitrypsin (650 μg/ml; 50 times higher than previous results using random insertion of the transgene).

The work on transgenic milk production is expensive requiring large facilities for purifying the protein from sheep's milk. Purification is important because even when 99.9% pure, human patients can develop antibodies against the tiny amounts of sheep proteins that remain.

GTC Bio-therapeutics, won preliminary approval to market a human protein, anti-thrombin, in Europe. Their protein, the first made in a transgenic animal to receive regulatory approval for human therapy and was secreted in the milk of transgenic goats.

Chickens

Chickens have several advantages over other farm animals:

1. Grow faster than sheep and goats and large numbers can be grown in close quarters

2. Synthesize several grams of protein in the "white" of their eggs.

Two methods have succeeded in producing chickens carrying and expressing foreign genes:

1. Infecting embryos with a viral vector carrying the human gene for a therapeutic protein and promoter sequences that will respond to the signals for making proteins (e.g., lysozyme) in egg white.

2. Transforming rooster sperm with a human gene and the appropriate promoters and checking for any transgenic offspring.

Preliminary results from both methods indicate that it may be possible for chickens to produce as much as 0.1 g of human protein in each egg that they lay. Not only should this cost less than producing therapeutic proteins in culture vessels, but chickens will probably add the correct sugars to glycosylated proteins something that E. coli cannot do.

Transgenic Fish

Aquatic animals are being engineered to increase aquaculture production, for medical and industrial research, and for ornamental reasons. Genes inserted to promote disease resistance may allow transgenic fish to absorb higher levels of toxic substances, including heavy metals. In turn, consumers of these fish may be ingesting higher amounts of substances such as mercury and selenium.

Transgenic fish that have genes from species such as peanuts or shellfish that are common causes of allergic reactions in humans may prompt allergic reactions in an unsuspecting consumer. Transgenic species may behave much like invasive species when interacting with the natural environment.

They may compete with native species for resources and pose a threat to the genetic diversity of native populations, especially when genetic modifications such as a rapid growth rate offer advantages over slower-developing native species. Despite industry assurances that transgenic fish would be unable to naturally reproduce or significantly threaten the environment, some scientists are far more doubtful.

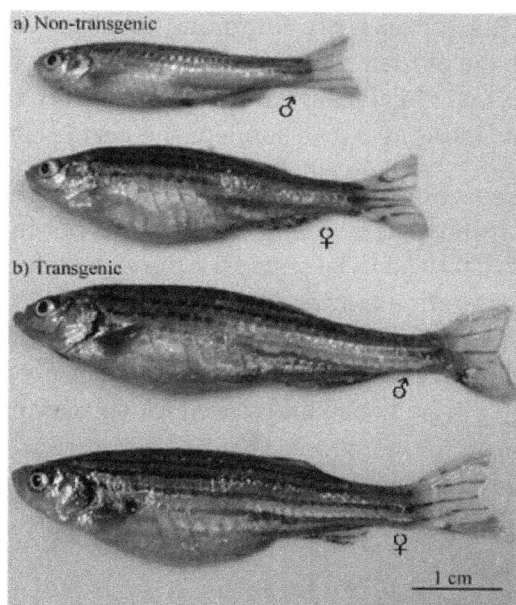

Transgenic fish.

The sample bill included in this package addresses these concerns by banning the importation, transportation, possession, spawning, incubation, cultivation, or release of aquatic transgenic animals except under a permit.

Fluorescent Cat

Recently South Korean scientist produced transgenic white Turkish angora cats to glow red under ultraviolet light these cats contain a fluorescent gene for flu protein and expressed under skin. Subsequently they produced a number of cloned cats from the skin cells of transformed mother cat. They proposed that such cat could be beneficial in diagnosis of genetic diseases and also showed a way to produce endangered animal by cloning.

Embryonic Stem Cell-mediated Gene Transfer

This method involves prior insertion of the desired DNA sequence by homologous recombination into an in vitro culture of embryonic stem (ES) cells. Stem cells are undifferentiated cells that have the potential to differentiate into any type of cell (somatic and germ cells) and therefore to give rise to a complete organism. These cells are then incorporated into an embryo at the blastocyst stage of development. The result is a chimeric animal. ES cell-mediated gene transfer is the method of choice for gene inactivation, the so-called knock-out method.

This technique is of particular importance for the study of the genetic control of developmental processes. This technique works particularly well in mice. It has the advantage of allowing precise targeting of defined mutations in the gene via homologous recombination.

Retrovirus-mediated Gene Transfer

To increase the probability of expression, gene transfer is mediated by means of a carrier or vector, generally a virus or a plasmid. Retroviruses are commonly used as vectors to transfer genetic material into the cell, taking advantage of their ability to infect host cells in this way. Offspring derived from this method are chimeric, i.e., not all cells carry the retrovirus. Transmission of the transgene is possible only if the retrovirus integrates into some of the germ cells.

For any of these techniques the success rate in terms of live birth of animals containing the transgene is extremely low. Providing that the genetic manipulation does not lead to abortion, the result is a first generation (F1) of animals that need to be tested for the expression of the transgene. Depending on the technique used, the F1 generation may result in chimeras. When the transgene has integrated into the germ cells, the so-called germ line chimeras are then inbred for 10 to 20 generations until homozygous transgenic animals are obtained and the transgene is present in every cell. At this stage embryos carrying the transgene can be frozen and stored for subsequent implantation.

Transgenic Animals in Human Welfare

The benefits of these animals to human welfare can be grouped into areas:

- Agriculture
- Medicine
- Industry

The examples below are not intended to be complete but only to provide a sampling of the benefits.

Agricultural Applications

There are many potential applications of transgenic methodology to develop new and improved strains of livestock. Practical applications of transgenics in livestock production include enhanced prolificacy and reproductive performance, increased feed utilization and growth rate, improved carcass composition, improved milk production and/or composition, modification of hair or fiber, and increased disease resistance. Development of transgenic farm animals will allow more flexibility in direct genetic manipulation of livestock. Gene transfer is a relatively rapid way of altering the genome of domestic livestock. The use of these tools will have a great impact toward improving the efficiency of livestock production and animal agriculture in a timely and more cost-effective manner. With ever-increasing world population and changing climate conditions, such effective means of increasing food production are needed.

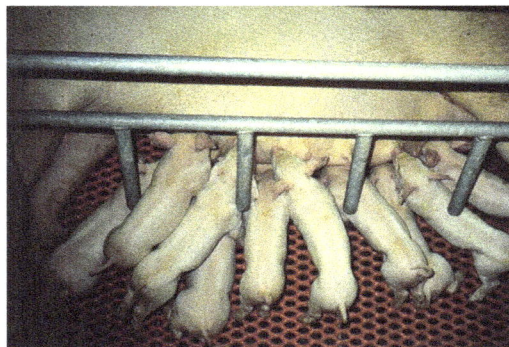

A litter of α-LA transgenic piglets. Transgenic sows produce up to 70% more milk than control non-transgenic litter mates. Piglets grow.

Enhanced Nutrition

Human health is directly affected by the necessity for a sustainable and secure supply of healthful food. Genetic modification of livestock holds the promise to improve public health via enhanced nutrition. For thousands of years, farmers have improved livestock in order to provide for nutritious, wholesome, and cost-effective animal products.

Transgenesis allows improvement of nutrients in animal products, including their quantity, the quality of the whole food, and specific nutritional composition. Transgenic technology could provide a means of transferring or increasing nutritionally beneficial traits. For example, enhancing the omega-3 fatty acid in fish consumed by humans may contribute to a decreased occurrence of coronary heart disease. In fact, transgenic pigs that contain elevated levels of omega-3 fatty acids have been produced. Furthermore, transfer of a transgene that elevates the levels of omega-3 fatty acids into pigs may enhance the nutritional quality of pork (Lai et al. 2006). The production of lower fat, more nutritious animal products produced by transgenesis could enable improvements in public health.

Reduced Environmental Impact

Over the last few years, livestock production has been under attack as being harmful to the environment. However, the production of transgenic livestock has the potential to dramatically reduce the environmental footprint of animal agriculture. Increasing efficiency and productivity through transgenesis could decrease the use of limited land and water resources while protecting the soil and ground water. One excellent example of this is the swine (the Enviro-PigTM) produced by genetic engineering. Pigs do not fully utilize dietary phosphorus. Dietary supplementation results in increased production costs, and incomplete utilization results in phosphorus levels in waste products that can cause pollution problems. Golovan et al. reported the production of transgenic pigs expressing salivary phytase as early as 7 d of age. The salivary phytase provided essentially complete digestion of dietary phytate phosphorus in addition to reducing phosphorus output by up to 75%. The use of phytase transgenic pigs in commercial pork production could result in decreased environmental phosphorus pollution from livestock operations.

Improved production efficiencies of milk and meat would decrease the amount of manure, slow the direct competition for human food, decrease the amount of water required for the animals and the production facilities, and decrease the land necessary for livestock operations.

Enhancing Milk

Advances in transgenic technology provide the opportunity either to change the composition of milk or to produce entirely novel proteins in milk. The improvement of livestock growth or survivability through the modification of milk composition involves production of transgenic animals that: (1) produce a greater quantity of milk; (2) produce milk of higher nutrient content; or (3) produce milk that contains a beneficial 'nutriceutical' protein. The major nutrients in milk are protein, fat, and lactose. By elevating any of these components, we can impact the growth and health of the developing offspring. Cattle, sheep, and goats used for meat production can benefit from increased milk yield or composition. In tropical climates, heat-tolerant livestock breeds such as Bos indicus cattle are essential for the expansion of agricultural production. However, Bos indicus cattle breeds do not produce copious quantities of milk. Improvement in milk yield by as little as 2-4 liters per day may have a profound effect on weaning weights in cattle such as the Nelore or Guzerat breeds in Brazil. Similar comparisons can be made with improving weaning weights in meat-type breeds like the Texel sheep and Boer goat. This application of transgenic technology could lead to improved growth and survival of offspring.

Table: Mammary expression of transgenic proteins.

Protein Expressed	Species Where Expressed	Promoter
Lysozyme	Goat	Bovine αs1-casein
Lysostaphin	Cattle	Ovine β-lactoglobulin
Bovine β and κ casein	Cattle	Bovine β-casein
IGF-I	Pig	Bovine α-lactalbumin
α-lactalbumin	Pig	Bovine α-lactalbumin
IGF-I	Rabbits	Bovine αs1-casein
Lactoferrin	Cattle	Bovine αs1-casein

The overexpression of beneficial proteins in milk through the use of transgenic animals may improve growth, development, health, and survivability of the developing offspring. Some factors that have been suggested to have important biological functions in the neonate that are obtained through milk include IGF-I, EGF, TGF-β, and lactoferrin.

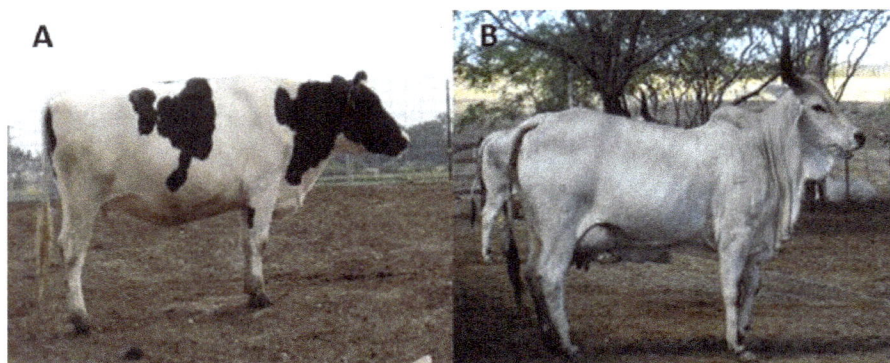

Can transgenic technology produce comparable milk volume? Small improvements in milk volume in Guzerat cows (left) using genetic material from high-producing Holsteins (right) could have a significant impact on Brazilian beef production.

Enhancing Growth Rates and Carcass Composition

The production of transgenic livestock has been instrumental in providing new insights into the mechanisms of gene action implicated in the control of growth. It is possible to manipulate growth factors, growth factor receptors, and growth modulators through the use of transgenic technology. Results from one study have shown that an increase in porcine growth hormone (GH) leads to enhancement of growth and feed efficiency in pigs. In the case of fish, there is a need for more efficient and rapid production, without diminishing the wild stocks, to provide a protein source for the increasing world population. The production of GH transgenic fish has led to dramatic (30-40%) increases in growth rates in catfish through the introduction of salmon GH into these animals.

Introduction of salmon GH constructs has resulted in a 5-11 fold increase in weight after 1 year of growth. This illustrates the point that increased growth rate and ultimately increased protein production per animal can be achieved via transgenic methodology.

Another aspect of manipulating carcass composition is that of altering the fat or cholesterol composition of the carcass. By altering the metabolism or uptake of cholesterol and/or fatty acids, the content of fat and cholesterol of meats, eggs, and cheeses could be lowered. There is also the possibility of introducing beneficial fats such as the omega-3 fatty acids from fish or other animals into our livestock. In addition, receptors such as the low-density lipoprotein (LDL) receptor gene and hormones like leptin are potential targets that would decrease fat and cholesterol in animal products.

Enhanced Animal Welfare through Improved Disease Resistance

Genetic modification of livestock will enhance animal welfare by producing healthier animals. Animal welfare is a high priority for anyone involved in the production of livestock. The application of transgenic methodology should provide opportunities to genetically engineer livestock with superior disease resistance.

One application of this technology is to treat mastitis, an inflammation of the mammary gland, typically caused by infectious pathogen(s). Mastitis causes decreased milk production. Transgenic dairy cows that secrete lysostaphin into their milk have higher resistance to mastitis due to the protection provided by lysostaphin, which kills the bacteria Staphylococcus aureus, in a dose dependent manner (Donovan et al. 2005). Lysostaphin is an antimicrobial peptide that protects the mammary gland against this major mastitis-causing pathogen.

Recent progress has produced prion-free and suppressed prion livestock. Prions are the causative agents in bovine spongiform encephalopathy (BSE) or 'mad cow disease' in cattle and in Creutzfeldt-Jacob disease (CJD) in humans. This is only a partial list of organisms or genetic diseases that decrease production efficiency and may also be targets for manipulation via transgenic methodologies.

Improving Reproductive Performance and Fecundity

Several potential genes have recently been identified that may profoundly affect reproductive performance and prolificacy. Introduction of a mutated or engineered estrogen receptor (ESR) gene could increase litter size in a number of diverse breeds of pigs. A single major autosomal gene for fecundity, the Boroola fecundity (FecB) gene, which allows for increased ovulation rate, has been identified in Merino sheep. Each copy of the gene has been shown to increase ovulation rate by approximately 1.5 ova. Production of transgenic sheep containing the appropriate FecB allele could increase fecundity in a number of diverse breeds. The manipulation of reproductive processes using transgenic methodologies is only beginning and should be a rich area for investigation in the future.

Improving Hair and Fiber

The control of the quality, color, yield, and even ease of harvest of hair, wool, and fiber for fabric and yarn production has been another area of focus for transgenic manipulation in livestock. The manipulation of the quality, length, strength, fineness, and crimp of the wool and hair fiber from sheep and goats has been examined using transgenic methods. In the future, transgenic manipulation of wool will focus on the surface of the fibers. Decreasing the surface interaction could decrease shrinkage of garments made from such fibers.

Recently, a novel approach to producing spider silk, a useful fiber, has been accomplished using the milk of transgenic goats. Spiders that produce orb-webs synthesize as many as seven different types of silk for making these webs. One of the most durable varieties is dragline silk. This material can be elongated up to 35% and has tensile properties close to those of the synthetic fiber KevlarTM. Its energy-absorbing capabilities exceed those of steel. There are numerous potential applications of these fibers in medical devices, sutures, ballistic protection, tire cord, air bags, aircraft, automotive composites, and clothing.

Medical Applications

Transplant organs may soon come from transgenic animals.

a) Xenotransplantation:

Patients die every year for lack of a replacement heart, liver, or kidney. For example, about 5,000 organs are needed each year in the United Kingdom alone. Transgenic pigs may provide the transplant organs needed to alleviate the shortfall. Currently, xenotransplantation is hampered by a pig protein that can cause donor rejection but research is underway to remove the pig protein and replace it with a human protein.

Milk-producing transgenic animals are especially useful for medicines.

b) Nutritional supplements and pharmaceuticals:

Products such as insulin, growth hormone, and blood anti-clotting factors may soon be or have already been obtained from the milk of transgenic cows, sheep, or goats. Research is also underway to manufacture milk through transgenesis for treatment of debilitating diseases such as phenylketonuria (PKU), hereditary emphysema, and cystic fibrosis.

In 1997, the first transgenic cow, Rosie, produced human protein-enriched milk at 2.4 grams per litre. This transgenic milk is a more nutritionally balanced product than natural bovine milk and could be given to babies or the elderly with special nutritional or digestive needs. Rosie's milk contains the human gene alpha-lactalbumin.

A transgenic cow exists that produces a substance to help human red cells grow.

c) Human gene therapy:

Human gene therapy involves adding a normal copy of a gene (transgene) to the genome of a person carrying defective copies of the gene. The potential for treatments for the 5,000 named genetic diseases is huge and transgenic animals could play a role. For example, the A. I. Virtanen Institute in Finland produced a calf with a gene that makes the substance that promotes the growth of red cells in humans.

Uses in industry include material fabrication and safety tests of chemicals.

Industrial Applications

In 2001, two scientists at Nexia Biotechnologies in Canada spliced spider genes into the cells of lactating goats. The goats began to manufacture silk along with their milk and secrete tiny silk strands from their body by the bucketful. By extracting polymer strands from the milk and weaving them into thread, the scientists can create a light, tough, flexible material that could be used in such applications as military uniforms, medical microsutures, and tennis racket strings.

Toxicity-sensitive transgenic animals have been produced for chemical safety testing. Microorganisms have been engineered to produce a wide variety of proteins, which in turn can produce enzymes that can speed up industrial chemical reactions.

References

- Recent-trends-in-animal-biotechnology, animal-biotechnology, biotechnology: biologydiscussion.com, Retrieved 21 March, 2019

- Margawati, biotechnology: actionbioscience.org, Retrieved 12 April, 2019

- Transgenic-animals-an-overview, transgenic-animals, animals: biologydiscussion.com, Retrieved 15 May, 2019

- Transgenic, browder: ucalgary.ca, Retrieved 23 June, 2019

- Production-of-transgenic-animals-using-virus-as-a-vector, transgenic-animals, animals: biologydiscussion.com, Retrieved 20 July, 2019

- Margawati, biotechnology: actionbioscience.org, Retrieved 25 August, 2019

- Transgenic-animals-in-agriculture, library, knowledge, scitable: nature.com, Retrieved 31 July, 2019

- Margawati, biotechnology: actionbioscience.org, Retrieved 17 February, 2019

6

Environmental Biotechnology

The branch of biotechnology which uses biological processes to tackle environmental problems such as the removal of pollution and biomass production is known as environmental biotechnology. Bioremediation is an example of environmental biotechnology where microorganisms are used to consume and break down environmental pollutants. This chapter discusses in detail the theories and methodologies related to environmental biotechnology.

Environmental biotechnology in particular is the application of processes for the protection and restoration of the quality of the environment. Environmental biotechnology can be used to detect, prevent and remediate the emission of pollutants into the environment in a number of ways.

Solid, liquid and gaseous wastes can be modified, either by recycling to make new products, or by purifying so that the end product is less harmful to the environment. Replacing chemical materials and processes with biological technologies can reduce environmental damage.

In this way environmental biotechnology can make a significant contribution to sustainable development. Environmental Biotechnology is one of today's fastest growing and most practically useful scientific fields. Research into the genetics, biochemistry and physiology of exploitable microorganisms is rapidly being translated into commercially available technologies for reversing and preventing further deterioration of the earth's environment.

Objectives of Environmental Biotechnology

The aim of environmental biotechnology is to prevent, arrest and reverse environmental degradation through the appropriate use of biotechnology in combination with other technologies, while supporting safety procedures as a primary component of the programme.

Specific objectives are:

1. To adopt production processes that make optimal use of natural resources, by recycling biomass, recovering energy and minimizing waste generation.

2. To promote the use of biotechnological techniques with emphasis on bioremediation of land and water, waste treatment, soil conservation, reforestation, afforestation and land rehabilitation.

3. To apply biotechnological processes and their products to protect environmental integrity with a view to long-term ecological security.

Use of biotechnology to treat pollution problems is not a new idea. Communities have depended on complex populations of naturally occurring microbes for sewage treatment for over a century. Every living organism—animals, plants, bacteria and so forth—ingests nutrients to live and produces a waste as a by-product. Different organisms need different types of nutrients.

Certain bacteria thrive on the chemical components of waste products. Some microorganisms feed on materials toxic to others. Research related environmental biotechnology is vital in developing effective solutions for mitigating, preventing and reversing environmental damage with the help of these living forms. Growing concern about public health and the deteriorating quality of the environment has prompted the development of a range of new, rapid analytical devices for the detection of hazardous compounds in air, water and land. Recombinant DNA technology has provided the possibilities for the prevention of pollution and holds a promise for a further development of bioremediation.

Applications of Environmental Biotechnology

Environmental protection is an integral component of sustainable development. The environment is threatened every day by the activities of man. With the continued increase in the use of chemicals, energy and non-renewable resources by an expanding global population, associated environmental problems are also increasing. Despite escalating efforts to prevent waste accumulation and to promote recycling, the amount of environmental damage caused by over-consumption, the quantities of waste generated and the degree of unsustainable land use appear likely to continue growing.

The remedy can be achieved, to some extent, by the application of environmental biotechnology techniques, which use living organisms in hazardous waste treatment and pollution control. Environmental biotechnology includes a broad range of applications such as bioremediation, prevention, detection and monitoring, genetic engineering for sustainable development and better quality of living.

Bioremediation

Bioremediation refers to the productive use of microorganisms to remove or detoxify pollutants, usually as contaminants of soils, water or sediments that otherwise intimidate human health. Bio treatment, bio reclamation and bio restoration are the other terminologies for bioremediation. Bioremediation is not a new practice. Microorganisms have been used for many years to remove organic matter and toxic chemicals from domestic and manufacturing waste discharge.

However, the focus in environmental biotechnology for fighting different pollution is on bioremediation. The vast majority of bioremediation applications use naturally occurring microorganisms to identify and filter toxic waste before it is introduced into the environment or to clean up existing pollution problems.

Some more advanced systems using genetically modified microorganisms are being tested in waste treatment and pollution control to remove difficult-to-degrade materials. Bioremediation can be performed in situ or in specialized reactors (ex situ). Bioremediation by microorganisms need appropriate environment for the clean up of the polluted site.

Addition of nutrients, terminal electron acceptors (O_2/NO_2), temperature, moisture to promote the growth of a particular organism may be required for the microbial activity in the polluted site. Bioremediation operations may be made either on-site or off-site, in situ or ex situ. Bioremediation has a vast potential to clean up water and soil contaminated by a variety of hazardous pollutants, domestic wastes, radioactive wastes etc.

Biological cleaning procedures make use of the fact that most organic chemicals are subjected to enzymatic attack of living organisms. The most common approach is the use of enzymes as substitute chemical catalysts. Significant reduction or complete elimination of harsh chemicals may be achieved as is observed in leather, textile processing and pulp and paper industry.

Only 1-2g of hemicellulose is substituted for 10-15 kg of chlorine to treat 1 tonne of pulp, thereby significantly reducing the chlorinated organic effluent. Environmental protection and remediation presently combine biotechnological, chemical, physical and engineering methods.

The relative importance of biotechnology is increasing as scientific knowledge and methods improve. Its lower requirements for energy and chemicals, combined with lower production of minor wastes, make it an increasingly desirable alternative to more traditional chemical and physical methods of remediation. Applications of bioremediation for maintenance of environment are several.

Waste Water and Industrial Effluents

Water pollution is a serious problem in many countries of the world. Rapid industrialisation and urbanization have generated large quantities of waste water that resulted in deterioration of surface water resources and ground water reserves. Biological, organic and inorganic pollutants contaminate the water bodies.

In many cases, these sources have been rendered unsafe for human consumption as well as for other activities such as irrigation and industrial needs. This illustrates that degraded water quality can, in effect, contribute to water scarcity as it limits its availability for both human use and the ecosystem. Treatment of the waste water before disposal is of urgent concern worldwide.

In sewage treatment plants microorganisms are used to remove the more common pollutants from waste water before it is discharged into rivers or the sea. Increasing industrial and agricultural pollution has led to a greater need for processes that remove specific pollutants such as nitrogen and phosphorus compounds, heavy metals and chlorinated compounds.

Methods include aerobic, anaerobic and physico-chemical processes in fixed-bed filters and in bioreactors in which the materials and microbes are held in suspension. Sewage and other waste waters would, if left untreated, undergo self-purification but the process requires long exposure periods. To speed up this process bioremediation measures are used.

However, Five Key Stages are Recognized in Wastewater Treatment:

a) Preliminary treatment – grit, heavy metals and floating debris are removed.

b) Primary treatment – suspended matters are removed.

c) Secondary treatment – bio-oxidize organic materials by activities of aerobic and anaerobic microorganisms.

d) Tertiary treatment – specific pollutants are removed (ammonia and phosphate).

e) Sludge treatment – solids are removed (final stage).

Aerobic Biological Treatment

Trickling filters, rotating biological contactors or contact beds, usually consist of an inert material (rocks/ash/wood/metal) on which the microorganisms grow in the form of a complex biofilm. These have been used for more than 70 years for sewage and waste water treatment. In these processes the degradable organic matter is oxidized by the microorganisms to CO_2 that can be vented to the atmosphere.

Activated Sludge Process

This process is used for treatment and removal of dissolved and biodegradable wastes, such as organic chemicals, petroleum refining wastes textile wastes and municipal sewage. The microorganisms in activated sludge generally are composed of 70-90% organic and 10-30% inorganic matters.

The microorganisms found in this sludge are usually bacteria, fungi, protozoa and rotifers. Petroleum hydrocarbons are degraded by species of bacteria (Acinetobacter, Mycobacteria, Pseudomonas etc.), yeasts, Cladosporium and Scolecobasidium. Pesticides (aldrin, dieldrin, parathion, malathion) are detoxified by fungus Xylaria xylestrix. Pseudomonas (a predominant soil micrganism) can detoxify organic compounds like hydrocarbons, phenols, organophosphates, polychlorinated biphenyls and polycyclic aromatics.

Utilisation of immobilized cyanobacterium Phormidium laminosum in batch and continuous flow bioreactors for the removal of nitrate, nitrite and phosphate from water has been reported by Garbisu et al. Blanco et al. showed the biosorption of heavy metal by Phormidium laminosum immobilised in micro-porous polymeric matrices. Photo-bioreactors are currently used to grow algae and cyanobacteria under closely controlled environmental conditions, with a view to making high-value products (such as beta-carotene and gamma-linoleic acid), designing efficient effluent treatment processes, and providing new energy sources.

The costs of wastewater treatment can be reduced by the conversion of wastes into useful products. Sulphur metabolizing bacteria can remove heavy metals and sulphur compounds from waste streams of the galvanization industry and reused. Most anaerobic wastewater treatment systems produce useful biogas.

In some cases, the by-products of the pollution-fighting microorganisms are themselves useful. Methane, for example, can be derived from a form of bacteria that degrades sulphur liquor, a waste product of paper manufacturing.

Soil and Land Treatment

As the human population grows, its demand for food from crops increases, making soil conservation crucial. Deforestation, over-development, and pollution from man-made chemicals are just a few of the consequences of human activity and carelessness. The increasing amounts of fertilizers and other agricultural chemicals applied to soils and industrial and domestic waste-disposal practices, led to the increasing concern of soil pollution. Pollution in soil is caused by persistent toxic compounds, chemicals, salts, radioactive materials, or disease-causing agents, which have adverse effects on plant growth and animal health.

Many species of fungi can be used for soil bioremediation. Lipomyces sp. can degrade herbicide paraquat. Rhodotorula sp. can convert benzaldehyde to benzyl alcohol. Candida sp. degrades formaldehyde in the soil. Aspergillus niger and Chaetomium cupreum are used to degrade tannins (found in tannery effluents) in the soil thereby helping in plant growth.

Phanerochaete chrysosporium has been used in bioremediation of soils polluted with different chemical compounds, usually recalcitrant and regarded as environmental pollutants. Decrease of PCP (Pentachlorophenol) between 88-91% within six weeks was observed in presence of Phanerochaete chrysosporium.

Bioremediation of contaminated soil has been used as a safe, reliable, cost-effective and environment friendly method for degradation of various pollutants. This can be effected in a number of ways, either in situ or by mechanically removing the soil for treatment elsewhere.

In situ treatments include adding nutrient solutions, introducing microorganisms and ventilation. Ex situ treatment involves excavating the soil and treating it above ground, either as compost, in soil banks, or in specialised slurry bioreactors. Bioremediation of land is often cheaper than physical methods and its products are largely harmless.

During biological treatment soil microorganisms convert organic pollutants to CO2, water and biomass. Degradation can take place under aerobic as well as under anaerobic conditions. Soil bioremediation can also be accomplished with the help of bioreactors. Degradation can take place under aerobic as well as under anaerobic conditions. Soil bioremediation can also be accomplished with the help of bioreactors. Liquids, vapours, or solids in a slurry phase are treated in a reactor. Microbes can be of natural origin, cultivated or even genetically engineered.

Research in the field of environmental biotechnology has made it possible to treat soil contaminated with mineral oils. Solid-phase technologies are used for petroleum-contaminated soils that are excavated, placed in a containment system through which water and nutrients percolate. Biological degradation of oils has proved commercially viable both on large and small scales, in situ and ex situ.

In situ soil bioremediation involve the stimulation of indigenous microbial populations (e.g. by adding nutrients or aeration). In this process the environmental conditions for the biological degradation of organic pollutants are optimized as far as possible. Oxygen has to be supplied by artificial aeration or by adding electron acceptors such as nitrates or oxygen releasing compounds. Ozone dissolved in water and H_2O_2 are sometimes used which degrade the organic contaminants.

Air and Waste Gases

With the onset of human civilization, the air is one of the first and most polluted components of the atmosphere. Most air pollution comes from one human activity: burning fossil fuels—natural gas, coal, and oil—to power industrial processes and motor vehicles. When fuels are incompletely burned, various chemicals called volatile organic chemicals (VOCs) also enter the air. Pollutants also come from other sources.

For instance, decomposing garbage in landfills and solid waste disposal sites emits methane gas, and many household products give off VOCs. Expanding industrial activities have added more contaminants in the air.

The concept of biological air treatment at first seemed impossible. With the development of biological waste gas purification technology using bioreactors—which includes bio filters, bio trickling filters, bio scrubbers and membrane bioreactors—this problem is taken care of. The mode of operation of all these reactors is similar.

Air containing volatile compounds is passed through the bioreactors, where the volatile compounds are transferred from the gas phase into the liquid phase. Microbial community (mixture of different bacteria, fungi and protozoa) grow in this liquid phase and remove the compounds acquired from the air.

In the bio filters, the air is passed through a bed packed with organic materials that supplies the necessary nutrients for the growth of the microorganisms. This medium is kept damp by maintaining the humidity of the incoming air. Biological off-gas treatment is generally based on the absorption of the VOC in the waste gases into the aqueous phase followed by direct oxidation by a wide range of voracious bacteria, which include Nocardia sp. and Xanthomonas sp.

Prevention

Sustainable development and quality living depends upon the rational, eco-friendly use of natural resources with economic growth. To comply with this trend, industrial development has to change to sustainable style from degradative type and for such a purpose cleaner technologies have to be adopted.

According to United Nations Environment Programme 'the continuous application of an integrated preventive environmental strategy to processes, products and services to increase eco-efficiency and reduce risks to humans and the environment' defines the eco-friendly concept. The application of preventive and clean concept can only be achieved by the 5R policies.

Five Environmental Buzzwords are the 5Rs for Efficient Use of Energy and Better Control of Waste, Which Might Help in Sustainable Development and Quality Living:

1. Reduce (Reduction of waste)

2. Reuse (Efficient use of water, energy)

3. Recycle (Recycling of wastes)

4. Replace (Replacement of toxic/hazardous raw materials for more environment- friendly inputs)

5. Recover (useful non-toxic fractions from wastes)

Innovation and adoption of clean technologies is the target of research and development worldwide. Industrial companies are developing processes with reduced environmental impact responding to the international call for the development of a sustainable society. There is a pervading trend towards less harmful products and processes; away from "end-of-pipe" treatment of waste streams. Environmental biotechnology, with its appropriate technologies, is suitable to contribute to this trend.

Enzyme Application

Enzymes are widely employed in industries for many years. Enzymes, non-toxic and biodegradable, are biological catalysts that are highly competent and have numerous advantages over non-biological catalysts. The use of enzyme by man, both directly and indirectly, have been for thousands of years.

In the recent years enzymes have played important roles in the production of drugs, fine chemicals, amino acids, antibiotics and steroids. Industrial processes can be made eco-friendly by the use of enzymes. Enzyme application in the textile, leather, food, pulp and paper industries help in significant reduction or complete elimination of severe chemicals and are also more economic in energy and resource consumption.

Biotechnological methods can produce food materials with improved nutritional value, functional characteristics, shelf stability. Plant cells grown in fermenters can produce flavours such as vanilla, reducing the need for extracting the compounds from vanilla beans. Food processing has benefited from biotechnologically produced chymosin which is used in cheese manufacture; alpha-amylase, which is used in production of high-fructose corn syrup and dry beer; and lactase, which is added to milk to reduce the lactose content for persons with lactose intolerance.

Genetically engineered enzymes are easier to produce than enzymes isolated from original sources and are favoured over chemically synthesized substances because they do not create by-products or off-flavours in foods.

Environmental Detection and Monitoring

A wide range of biological methods are in use to detect pollution and for the continuous monitoring of pollutants. The techniques of biotechnology have novel methods for diagnosing environmental problems and assessing normal environmental conditions so that human beings can be better- informed of the surroundings. Applications of these methods are cheaper, faster and also portable.

Rather than gathering soil samples and sending them to a laboratory for analysis, scientists can measure the level of contamination on site and know the results immediately. Biological detection methods using biosensors and immunoassays have been developed and are now in the market. Microbes are used in biosensors contamination of metals or pollutants. Saccharomyces cerevisiae (yeast) is used to detect cyanide in river water while Selenastrum capricornatum (green alga) is used for heavy metal detection. Immunoassays use labelled antibodies (complex proteins produced in biological response to specific agents) and enzymes to measure pollutant levels. If a pollutant is present, the antibody attaches itself to it making it detectable either through colour change, fluorescence or radioactivity.

Biosensors

A biosensor is an analytical device that converts a biological response into an physical, chemical or electrical signal. The development of biosensors involves integration of a specific and sensitive biologically derived sensing elements (immobilized cells, enzymes or antibodies) are integrated with physico-chemical transducers (either electrochemical or optical). Immobilised on a substrate, their properties change in response to some environmental effect in a way that is electronically or optically detectable.

It is then possible to make quantitative measurements of pollutants with extreme precision or to very high sensitivities. The biological response of the biosensor is determined by the bio catalytic membrane, which accomplishes the conversion of reactant to product. Immobilised enzymes possess a number of advantageous features which makes them particularly applicable for use in such systems.

They may be re-used, which ensures that the same catalytic activity is present for a series of analyses. Biosensors are powerful tools, which rely on biochemical reactions to detect specific substances, which have brought benefits to a wide range of sectors, including the manufacturing, engineering, chemical, water, food and beverage industries. They are able to detect even small amounts of their particular target chemicals, quickly, easily and accurately.

For this character of biosensors they have been ardently adopted for a variety of process monitoring applications, principally in respect to pollution assessment and control. Biosensors for detection of carbohydrates, organic acids, glucosinolates, aromatic hydrocarbons, pesticides, pathogenic bacteria and others have already been developed.

The biosensors can be designed to be very selective, or sensitive to a broad range of compounds. For example, a wide range of herbicides can be detected in river water using algal-based biosensors; the stresses inflicted on the organisms being measured as changes in the optical properties of the plant's chlorophyll. Biosensors are of different types such as calorimetric biosensors, immunosensors, optical biosensors, BOD biosensors, gas biosensors.

The remarkable ability of microbes to break down chemicals is proving useful, not only in pollution remediation but also in pollutant detection. A group of scientists at Los Alamos National Laboratory work with bacteria that degrade a class of organic chemicals called phenols. When the bacteria ingest phenolic compounds, the phenols attach to a receptor.

The phenol-receptor complex then binds to DNA, activating the genes involved in degrading phenol. The Los Alamos scientists added a reporter gene that, when triggered by a phenol-receptor complex, produces an easily detectable protein, thus indicating the presence of phenolic compounds in the environment. Biosensors employing acetylcholine esterase can be used for the detection of organophosphorus compounds in water.

Different Areas of Environmental Biotechnology

Environmental Biotechnology and Metagenomics

Environmental Biotechnology is Divided into Different Areas:

- Direct studies of the environment,

- Research with a focus on applications to the environment and

- Research that applies information from the environment to other venues.

Here, a brief account of a particular aspect of direct analysis of environment is given.

In addition to DNA inside living organisms, there is much free DNA in the environment that might also be a source of new genes. The field of environmental biotechnology has revolutionized the study of the life-forms which have not been studied earlier and DNA.

This approach is direct analyses of the environment and the natural biochemical processes that are present. A significant study in this aspect is metagenomics. Metagenomics is the study of the genomes of whole communities of microscopic life forms and it deals with a mixture of DNA from multiple organisms, viruses, viroids, plasmids and free DNA.

In other words, metagenomics, the genomic analysis of a population of microorganisms, is the method to gain access to the physiology and genetics of uncultured organisms.

Using metagenomics, researchers investigate, catalogue the current microbial diversity. New proteins, enzymes and biochemical pathways are identified. The knowledge garnered from metagenomics has the potential to affect the ways we use the environment. Metagenomic analyses involves isolating DNA from an environmental sample, cloning the DNA into a suitable vector, transforming the clones into a host bacterium and screening the resultant transformants.

The clones can be screened for phylogenetic markers such as 16S rRNA and rec A or for other conserved genes by hybridization or multiplex PCR or for expression of specific traits such as enzyme activity or antibiotic production or they can be sequenced randomly.

One very important method for metagenomic study is stable isotope probing (SIP). An environmental sample of water or soil is first mixed with a precursor such as methanol, phenol, carbonate or ammonia that has been labeled with a stable isotope such as 15N, 13C or 18O. If the organisms in the sample metabolize the precursor substrate, the stable isotope is incorporated into their genome.

When the DNA from the sample is isolated and separated by centrifugation, the genomes that incorporated the labeled substrate will be heavier and can be separated from the other DNA in the sample. The heavier DNA will migrate further in a cesium chloride gradient during centrifugation. The DNA can be used directly or cloned into vectors to make a metagenomic library. This technique is useful to find new organisms that can degrade contaminants such as phenol.

Microorganisms are crucial participants in cleaning up a large variety of hazardous substances/ chemicals by transforming them into forms that are harmless to people and environment. One very important example is given here. Gasoline is leaked into soil in every gas station in United States.

There is every possibility that gasoline will be mixed with ground water which is the prime source of drinking water. However, the dormant members of the soil microbial community are triggered to become active and degrade the harmful chemicals in gasoline.

Since gasoline is composed of hundreds of chemicals it takes a variety of microbes working together to degrade them all. When some bacteria cause a depletion of O_2 in ground water near a gasoline spill, other types of bacteria that can use nitrate for energy begin biodegrading the gasoline. Bacteria that use iron, manganese and sulfate follow.

All these microbial communities work together in a pattern to transform leaking gasoline into CO_2 and water. Metagenomic analysis may help us identify the particular community member and function needed to achieve the full chemical transformation that will keep our planet livable.

Bioremediation

Bioremediation is a branch of biotechnology which deals with the use of living organisms such as microbes and bacteria to remove contaminants, pollutants and toxins from soil and water. It can be used to clean up environmental problems like an oil spill or contaminated ground-water.

Bioremediation relies on stimulating the growth of certain microbes that use contaminants like oil, solvents, and pesticides as a source of food and energy. These microbes consume the contaminants, converting them into small amounts of water and harmless gases like carbon dioxide. Effective bioremediation needs a combination of the right temperature, nutrients, and food; otherwise, it may take much longer for the cleanup of contaminants. If conditions are not favorable for bioremediation, they can be improved by adding "amendments" to the environment, such as molasses, vegetable oil or simply air. These amendments create optimum conditions for microbes to flourish and complete the bioremediation process.

Bioremediation can either be done in situ, at the site of the contamination itself, or ex situ, at a location away from the site. Ex situ bioremediation may be necessary if the climate is too cold for microbe activity, or the soil is too dense for nutrients to be spread evenly. Ex situ bioremediation may require digging up the soil and cleaning it above ground, which may greatly increase the cost of the process.

The process of bioremediation can take anywhere from a few months to several years. The amount of time required depends on variables such as the size of the contaminated area, the concentration of contaminants, conditions such as temperature and soil density, and whether bioremediation will take place in situ or ex situ.

Types of Bioremediation

(a) Mycoremediation:

This is a type of Bioremediation; fungi are used for the process of decontamination. The use of fungal mycelia in bioremediation is called Mycoremediation. The role of the fungus in the ecosystem is to perform the work of braking down the organic substances into much smaller and simpler materials. The mycelium helps in braking down the substances and they secrete extracellur enzymes and acids that brakes lignin and cellulose; these are building blocks of plant fiber. The key function of Mycoremediation is to target the right fungal species for a specific pollutant.

b) Phytoremediation:

The direct use of the green plants and their microorganisms used to balance or decrease the contaminated soils, sludges, sediments, surface water or ground water is called Phytoremediation. As per the Ancient Greek term phyto means plant and remedian means restoring balance. This type of bioremediation explains a way of treating the environmental problems with the help of plants. The element of Phytoremediation consists of contaminated soil, water, and air which are polluted and the plants are able to contain and eliminate the metals, pesticides, solvents, explosives, crude oil.

C) Microbial Remediation:

The use of microorganisms to degrade organic contaminants and to bind the use of metals in less bioavaliable form is called Microbial Remediation. â€" Aerobic and Anaerobic conditions. When the microbes need oxygen to perform its process is in the case of aerobic condition; if they can ample amount of oxygen they â€ll be able to give maximum amount of water and carbon through the conversion of contaminants and toxins. In case of anaerobic conditions the microbes perform their work without the presence of oxygen the chemical compounds present in the soil helps the anaerobic to perform its duties efficiently.

Bioremediation Techniques used in Environmental Biotechnology

Bioremediation focuses on the former while it is recognized that at present biological treatment processes play a major role in preventing and reducing the organic and inorganic environmental contamination from the municipal, industrial and agricultural sectors.

Bioremediation is based on the premise that naturally occurring bacteria in an impacted environment develop the means to degrade or tolerate the presence of organic contaminats. This indigenous activity can be considered passive or intrinsic bioremediation. Intrinsic bioremediation has been recognized most frequently for ground water (saturated systems) contaminated with hydrocarbons.

Sometime a more active remediation system is used. Semi- passive bioremedation systems are defined as those in-situ treatments that induce favourable conditions for accelerated biodegradation. The excellent remedial approach is aggressive bioremediation. It may be defined as engineered bioremediation to produce optimal treatment in both in-situ and ex-situ systems. Its advantage is much shorter treatment period, alleviating any construction limitations sooner than with passive system.

Microorganisms are able to degrade a chemical because they are using it as an energy source to promote their own growth. According to Glazer and Nikaido, bioremedidation is defined as a spontaneous or managed process in which biological, especially microbiological, catalysis acts on pollutants and thereby remedies or eliminates environmental contamination. Bioremediation is currently being used to decrease the organic chemical waste content of ground water, soils, effluent from food processing industry and chemical plants, oily sludge from petroleum refineries and oil spills in the ocean.

Degradation of pesticides, insecticides and fungicides in-situ (in soil) by microbes Pseudomonas A3, P .putida, P .aeruginosa and Serratia marinorubra either singly or as a consortia has been successfully attempted. Some microbial strains such as S. marinorubra, Bacillus species YW and YDLK consortia have also been used for decolorization and degradation of industrial effluents and textile mill azo dyes and effluents.

Heavy metal detoxification and biosorption by employing the bacterium bacillus species YW has been found to be effective in reducing hexavalent chromium to its non-toxic trivalent form and the chromate resistance and reduction was found to be plasmid mediated process.

The removal of less toxic trivalent chromim through biosorption using the EPS Azotobacter species as the biomatrix has also been successfully attempted. Bioremediation techniques fall into four categories — in situ treatment, composting, land farming, and above – ground reactors.

In-situ Bioremediation Techniques

In-situ, in Latin, means "in the original place". Thus, in-situ bioremediation means bioremediation based on the degradative activities of endogenous microbial populations. In other words, it relies on the indigenous microbial flora of subsurface soils and ground water.

It depends on the premise that the microorganisms already present in a contaminated site have adapted to the organic chemical wastes found there and are able to degrade some or all of the components of these wastes. The degradation by these so-called adapted microorganisms goes on until some nutrient or electron acceptor attains a limiting concentration. In most cases, oxygen level is the limiting factor, but phosphate and nitrate may also become limiting factor.

The stimulation of natural biotransformation by nutrient addition (N, P) to the environment is called enhanced in-situ bioremediation. The addition of nutrients to contaminated soils and also to hydrocarbon- contaminated marine environments has been investigated.

One of the best examples of enhanced in-situ bioremediation was the clean-up of the Exxon Valdez oil spill in Alaska. The spill, about 11 million gallons of crude oil, severely affected 350 miles of shoreline in Bligh Reef in Prince William Sound. In this case, fertilizers were used to accelerate the removal of oil from the beaches, supplying extra nutrients that were in limiting concentrations.

A single application of inorganic fertilizer was shown to speed up the disappearance of oil by a factor of two to three over its rate of disappearance on untreated site. Samples of oil taken at the end of that time from surfaces of treated beaches showed changes in composition consistent with extensive biodegradation. Thus enhanced in-situ bioremediation offers several potential advantages in the elimination of hazardous wastes.

Besides Pseudomonas species, HCI and Raulstonia species have also been successfully used for oil degradation. Most of the Psedumonas species strains harboured a catabolic plasmid, which encodes the genes for hydrocarbon degradation. The biotransformation of this plasmid to various bacteria in natural soil and marine water has been carried out, indicating the horizointal transfer of catabolic genes from one bacterium to another, paving way to create "Superbugs" for bioremediation in differing and metamorphosing ecosystems.

Bio-venting is one of the most widely used methods of remediating soils contaminated by petroleum hydrocarbons. Bio-inventing supplies air to an unsaturated soil zone by using a combination of pumps and blowers that apply a vacuum to the target area, while continuously injecting low volumes of air.

In-situ bioremediation under anaerobic conditions may also be enhanced by providing electron acceptors such as sulphate or nitrate. However, in-situ methanogenic bio-remediation has not, so far, been investigated thoroughly, although it is recognised that degradation of organic pollutants in anaerobic microniches in soil, sediment and groundwater environments contributes significantly to overall in-situ bioremediation rates.

Ex-situ Bioremediation

Ex-situ bioremediation techniques are usually aerobic and involve treatment of contaminated soils or sediments using solid or slurry-phase systems. Solid-phase systems include compost heaps, land-farming (soil-treatment units) and engineered biopiles.

Composting

Compost is a mixture of soil, partially decayed plants, and sometimes manure and commercial fertilizer. It is very rich in microorganisms. It has long been used by farmers and gardners to make soils more fertile and to improve crop yields. Compost heaps consist of contaminated soil or sediment supplemented with composting material in order to enhance water and air-holding capacity and improve physical handling properties.

Composting techniques are used for remediation of highly contaminated sites and have proved successful for military sites contaminated with explosives such as 2, 4, 6-trinitrotoluene (TNT), hexahydro-1, 3, 5-trinitro-l, 3, 5-triazine (RDX) and N- methyl-1, 2, 4, 6-tetranitroaniline (Tetry1).

Biodegradation or biotransformation of over 90% of these explosives was achieved within 80 days in a compost pile maintained at 55°C. After 150 days, a starting concentration of 18,000 mg of explosives per kg of soil was reduced to 74 mg per kg.

Land-farming

Landfarming is used to dispose of oily sludges from petroleum refinery operations. The oily sludge from refinery wastes is mixed with soil and subjected to enhanced in-situ bioremediation. The sludges may be pretreated or not. Biological pretreatment of refinery effluents partially mineralizes the organic waste components; the residual solid waste (sludge) then has a high content of aromatic hydrocarbon compounds. In contrast, untreated settled solids, such as the solids from tank bottoms, contain high amounts of aliphatic hydrocarbons and silt.

The terrain of a landfarm must be flat to minimize runoff; the soil should be light and loamy for proper aeration; and a clay layer should underlie the porous surface soil to reduce the possibility of groundwater contamination through seepage. The optimal temperature range for biodegradation is 20-30°C. Inorganic fetilizer is applied to the site to provide fixed nitrogen, and phosphate and Ca CO_3 are added to raise the pH of the soil-waste mixture to about 7.8. For untreated sludge, maximal oil biodegradation rates in soil are achieved at a hydrocarbon load of 5-10% by weight – that is , 100-200 metric tons of hydrocarbon per ha.

In such landfarms, about 50-70% of the applied organic waste is degraded before the next batch of sludge is applied. However, a disadvantage of landfarming is that the process is quite slow. Moreover, the heavy metal constituents of the sludge gradually accumulate in the landfarm soil. As a result, a plot of land used intensively as a landfarm cannot be used later for grazing livestock or growing crops.

Above-ground Bioreactors

The above-ground bioreactors are based on the same technology as fermenters. There are four common types of bioreactors – the stirred- tank reactor, the bubble column, the air lift, and the packed bed (or fixed bed). They are used for the treatment of either excavated soil or groundwater containing high levels of contaminants. Contaminated soil is mixed with water and introduced into the reactor as a slurry.

Granulated charcoal, plastic spheres, glass beads, or diatomaceous earth provide a large surface area for microbial growth in bioreactors. The large surface area of the microbial biofilm that forms on such supports leads to a rapid rate of biodegradation. The microbial inoculum may be taken from activated sludge from a sewage treatment plant, an indigenous population at the contaminated site, or from a pure culture of appropriate microbes.

Such bioreactors can be used in series to accomplish different kinds of degaradation. For example, the first reactor can be operated in an anaerobic mode and its effluent transferred to a second reactor operated in an aerobic mode. Mineralization requires aerobic conditions whereas bio-trasformations such as dehalogenations of certain compounds require anaerobic conditions.

Environmental Factors Influencing In-situ Bioremediation

Both the rate and the extent of microbial remediation of organic contaminats in-situ are affected by a number of environmental factors, some of which may be mainpulated whereas others are difficult to modify within the contaminated site.

pH

Most laboratory based bioremediation studied have been carried out in about neutral pH since the majority of bacteria show optimal growth at this pH range. In many studies, adjustment of pH enhances the rate of biodegradation.

Temperature

Temperature directly affects the metabolism and growth of bacteria. The vast majority of in-situ bioremediation applications have been carried out under mesophilic conditions (between 20 and 40 °C). However, the thermophilic or thermotrophic species are capable of degrading a diverse range of organic compounds in wastes or wastewaters at higher temperature (60-70°C). A variety of techniques have been utilized to increase the temperature in in-situ soil remediation applications. All these studies indicated that even modest increase in temperature may significantly increase bioremediation rates.

Water Content

Water content in soils or sediments in an important factor affecting bioremediation rates. Microbes generally require water activity values (aw) of 0.9-1.0 in order to metabolize and grow. The majority of bacteria grow optimally at aw values in the upper limit of this range.

Soil Type

In general, in-situ bioremediation rates are enhanced when the soil is granular or porous. In-situ degradation rates are slowed down under unfavourable geological characteristics which include low permeability of soil, faractured rock, and water-logged or arid conditions.

Nutrient Availability

In almost all cases nutrient supplementation (addition of N,P) significantly increases bioremediation rates. It has shown very good results in hydrocarbon bioremediation of soils and ground waters when N and P levels have been shown to be limiting.

External Electron Availability

Many in-situ bioremediation techniques use some provision of oxygen supply to enhance aerobic respiratory breakdown of organic contaminats. Addition of hydrogen peroxide is used to introduce oxygen. Hydrogen peroxide is about 7 times more soluble in water than oxygen and its decomposition in soil yields 0.5 mol of O_2 per mol of H_2O_2 introduced to contaminated site ($_2H_2O_2 \rightarrow {}_2H_2O + O_2$).

Bioavailability of Organic Pollutants

This is an important factor governing the rate of in- situ bioremediation. Improvements in bioremediation rates have been achieved by the addition of bio-surfactants or synthetic detergents to the contaminated zone. Addition of biodegradable solvents, which assist in desorption and dissolution rates also exhibits increase in the biodegradation of the adsorbed pollutants (e.g. adsorption of PAHs by soil particles)

Cometabolism

It is a process whereby microbes involved in the metabolism of a growth promoting substrate also transform other organic contaminants (cosubstrates) that are not growth supporting if supplied as sole carbon and energy source. Cometabolic transformation of organic pollutants is an important process in both aerobic and anaerobic environments. Bacterial transformation of DDT, PCBs provide examples of both aerobic and anaerobic cometabolic biodegradation. Provision of a readily metabolizable substrate may also promote pollutant transformation by enhancing the growth of associated microbes involved in the overall microbial activity.

Gene Expression

The ability of indigenous microorganisms to degrade organic pollutants is dependent on expression of the genes encoding the required uptake and degradative enzyme systems. If the bioavailable concentration of a pollutant is too low, the expression of inducible operons may not occur.

Advantages of Bioremediation

Bioremediation has a number of advantages over other cleanup methods. As it only uses natural processes, it is a relatively green method that causes less damage to ecosystems. It often takes place underground, as amendments and microbes can be pumped underground to clean up contaminants in ground water and soil; therefore, it does not cause much disruption to nearby communities.

The process of bioremediation creates few harmful byproducts, since contaminants and pollutants are converted into water and harmless gases like carbon dioxide. Finally, bioremediations is cheaper than most cleanup methods, as it does not require a great deal of equipment or labor. As of 2012, bioremediation has been used to clean up more than 100 Superfund sites around the United States.

Biodegradation

Biodegradation or biological degradation is the phenomenon of biological transformation of organic compounds by living organisms, particularly the microorganisms.

Biodegradation basically involves the conversion of complex organic molecules to simpler (and mostly non-toxic) ones. The term biotransformation is used for incomplete biodegradation of organic compounds involving one or a few reactions. Biotransformation is employed for the synthesis of commercially important products by microorganisms.

Consortia of Microorganisms for Biodegradation

A particular strain of microorganism may degrade one or more compounds. Sometimes, for the degradation of a single compound, the synergetic action of a few microorganisms (i.e. a consortium or cocktail of microbes) may be more efficient. For instance, the insecticide parathion is more efficiently degraded by the combined action of Pseudomonas aeruginosa and Psudomonas stulzeri.

Co-metabolism in Biodegradation

In general, the metabolism (breakdown) of xenobiotics is not associated with any advantage to the microorganism. That is the pollutant chemical cannot serve as a source of carbon or energy for the organism. The term co-metabolism is often used to indicate the non-beneficial (to the microorganism) biochemical pathways concerned with the biodegradation of xenobiotics. However, co-metabolism depends on the presence of a suitable substrate for the microorganism. Such compounds are referred to co-substrates.

Factors Affecting Biodegradation

Several factors influence biodegradation. These include the chemical nature of the xenobiotic, the capability of the individual microorganism, nutrient and O_2 supply, temperature, pH and redox potential. Among these, the chemical nature of the substrate that has to be degraded is very important.

Some of the relevant features are given hereunder:

i. In general, aliphatic compounds are more easily degraded than aromatic ones.

ii. Presence of cyclic ring structures and length chains or branches decrease the efficiency of biodegradation.

iii. Water soluble compounds are more easily degraded.

iv. Molecular orientation of aromatic compounds influences biodegradation i.e. ortho > para > meta.

v. The presence of halogens (in aromatic compounds) inhibits biodegradation.

Besides the factors listed above, there are two recent developments to enhance the biodegradation by microorganisms.

Bio-stimulation

This is a process by which the microbial activity can be enhanced by increased supply of nutrients or by addition of certain stimulating agents (electron acceptors, surfactants).

Bio-augmentation

It is possible to increase biodegradation through manipulation of genes. More details on this genetic manipulation i.e. genetically engineered microorganisms (GEMs), are described later. Bio-augmentation can also be achieved by employing a consortium of microorganisms.

Enzyme Systems for Biodegradation

Several enzyme systems (with independent enzymes that work together) are in existence in the microorganisms for the degradation of xenobiotics. The genes coding for the enzymes of bio-degradative pathways may be present in the chromosomal DNA or more frequently on the plasmids. In certain microorganisms, the genes of both chromosome and plasmid contribute for the enzymes of biodegradation. The microorganism Pseudomonas occupies a special place in biodegradation.

A selected list of xenobiotics and the plasmids containing the genes for their degradation is given in table.

Xenobiotic	Name of plasmid in pseudomonas
Naphthalene	NAH
Xylene	SYL
Xylene and toluene	TOL, pWWO, XYL-K
Salicylate	SAL
Camphor	CAM
3-chlorobenzene	pAC25

Recalcitrant Xenobiotics

There are certain compounds that do not easily undergo biodegradation and therefore persist in the environment for a long period (sometimes in years). They are labeled as recalcitrant.

There may be several reasons for the resistance of xenobiotics to microbial degradation:

i.   They may chemically and biologically inert (highly stable).

ii. Lack of enzyme system in the microorganisms for biodegradation.

iii. They cannot enter the microorganisms being large molecules or lack of transport systems.

iv. The compounds may be highly toxic or result in the formation highly toxic products that kill microorganisms.

There are a large number of racalcitrant xenobiotic compounds e.g. chloroform, freons, insecticides (DDT, lindane), herbicides (dalapon) and synthetic polymers (plastics e.g. polystyrene, polyethylene, polyvinyl chlorine).

It takes about 4-5 years for the degradation of DDT (75-100%) in the soil. A group of microorganisms (Aspergillus flavus, Mucor aternans, Fusarium oxysporum and Trichoderma viride) are associated with the slow biodegradation of DDT.

Bio-magnification

The phenomenon of progressive increase in the concentration of a xenobiotic compound, as the substance is passed through the food chain is referred to as bio-magnification or bioaccumulation. For instance, the insecticide DDT is absorbed repeatedly by plants and microorganism.

When they are eaten by fish and birds, this pesticide being recalcitrant, accumulates, and enters the food chain. Thus, DDT may find its entry into various animals, including man. DDT affects the nervous systems, and it has been banned in some countries.

Types of Bioremediation

The most important aspect of environmental biotechnology is the effective management of hazardous and toxic pollutants (xenobiotics) by bioremediation. The environmental clean-up process through bioremediation can be achieved in two ways—in situ and ex situ bioremediation.

In Situ Bioremediation

In situ bioremediation involves a direct approach for the microbial degradation of xenobiotics at the sites of pollution (soil, ground water). Addition of adequate quantities of nutrients at the sites promotes microbial growth. When these microorganisms are exposed to xenobiotics (pollutants), they develop metabolic ability to degrade them.

The growth of the microorganisms and their ability to bring out biodegradation are dependent on the supply of essential nutrients (nitrogen, phosphorus etc.). In situ bioremediation has been successfully applied for clean-up of oil spillages, beaches etc. There are two types of in situ bioremediation-intrinsic and engineered.

Intrinsic Bioremediation

The inherent metabolic ability of the microorganisms to degrade certain pollutants is the intrinsic bioremediation. In fact, the microorganisms can be tested in the laboratory for their natural capability of biodegradation and appropriately utilized.

Engineered in Situ Bioremediation

The inherent ability of the microorganisms for bioremediation is generally slow and limited. However, by using suitable physicochemical means (good nutrient and O_2 supply, addition of electron acceptors, optimal temperature), the bioremediation process can be engineered for more efficient degradation of pollutants.

Advantages of in Situ Bioremediation:

1. Cost-effective, with minimal exposure to public or site personnel.

2. Sites of bioremediation remain minimally disrupted.

Disadvantages of in situ bioremediation:

  1. Very time consuming process.

  2. Sites are directly exposed to environmental factors (temperature, O_2 supply etc.).

  3. Microbial degrading ability varies seasonally.

Ex Situ Bioremediation

The waste or toxic materials can be collected from the polluted sites and the bioremediation with the requisite microorganisms (frequently a consortium of organisms) can be carried out at designed places. This process is certainly an improvement over in situ bioremediation, and has been successfully used at some places.

Advantages of ex situ bioremediation:

  1. Better controlled and more efficient process.

  2. Process can be improved by enrichment with desired microorganisms.

  3. Time required in short.

Disadvantages of ex situ bioremediation:

  1. Very costly process.

  2. Sites of pollution are highly disturbed.

  3. There may be disposal problem after the process is complete.

Biodegradation of Hydrocarbons

Hydrocarbon are mainly the pollutants from oil refineries and oil spills. These pollutants can be degraded by a consortium or cocktail of microorganisms e.g. Pseudomonas, Corynebacterium, Arthrobacter, Mycobacterium and Nocardia.

Biodegradation of Aliphatic Hydrocarbons

The uptake of aliphatic hydrocarbons is a slow process due to their low solubility in aqueous medium. Both aerobic and anaerobic processes are operative for the degradation of aliphatic hydrocarbons. For instance, unsaturated hydrocarbons are degraded in both anaerobic and aerobic environments, while saturated ones are degraded by aerobic process. Some aliphatic hydrocarbons which are reclacitrant to aerobic process are effectively degraded in anaerobic environment e.g. chlorinated aliphatic compounds (carbon tetrachloride, methyl chloride, vinyl chloride).

Biodegradation of Aromatic Hydrocarbons

Microbial degradation of aromatic hydrocarbons occurs through aerobic and anaerobic processes. The most important microorganism that participates in these processes is Pseudomonas.

The biodegradation of aromatic compounds basically involves the following sequence of reactions:

1. Removal of the side chains.

2. Opening of the benzene ring.

Most of the non-halogenated aromatic compounds undergo a series of reactions to produce catechol or protocatechuate. The bioremediation of toluene, L-mandelate, benzoate, benzene, phenol, anthracene, naphthalene, phenanthrene and salicylate to produce catechol is shown, depicts the bioremediation of quinate, p-hydroxymandelate, p-hydroxybenzoyl formate, p-toluate, benzoate and vanillate to produce protocatechuate.

Catechol and protocatechuate can undergo oxidative cleavage pathways. In ortho-cleavage pathway, catechol and protocatechuate form acetyl CoA, while in meta-cleavage pathway, they are converted to pyruvate and acetaldehyde. The degraded products of catechol and protocatechuate are readily metabolised by almost all the organisms.

Biodegradation of Pesticides and Herbicides

Pesticides and herbicides are regularly used to contain various plant diseases and improve the crop yield. In fact, they are a part of the modern agriculture, and have significantly contributed to green revolution. The common herbicides and pesticides are propanil (anilide), propham (carbamate), atrazine (triazine), picloram (pyridine), dichlorodiphenyl trichloroethane (DDT) monochloroacetate (MCA), monochloropropionate (MCPA) and glyphosate (organophosphate). Most of the pesticides and herbicides are toxic and are recalcitrant (resistant to biodegradation). Some of them are surfactants (active on the surface) and retained on the surface of leaves.

Biodegradation of Halogenated Aromatic Compounds

Most commonly used herbicides and pesticides are aromatic halogenated (predominantly chlorinated) compounds. The bio-degradative pathways of halogenated compounds are comparable with that described for the degradation of non-halogenated aromatic compounds. The rate of degradation of halogenated compounds is inversely related to the number of halogen atoms that are originally present on the target molecule i.e. compounds with higher number of halogens are less readily degraded.

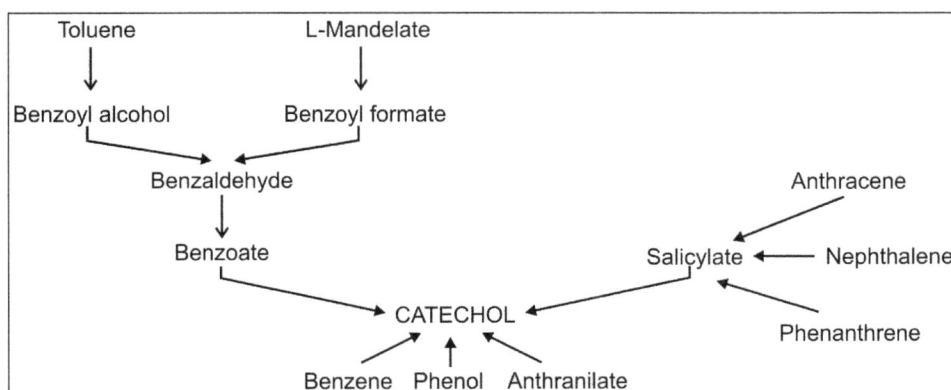

Bioremediation of Certain Aromatic compound by bacteria to produce catechol.

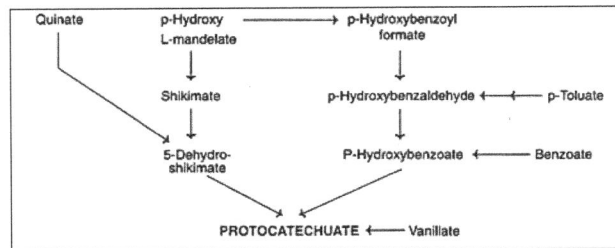

Bioremediation of Certain organic compounds by bacteria to produce protocatechuate.

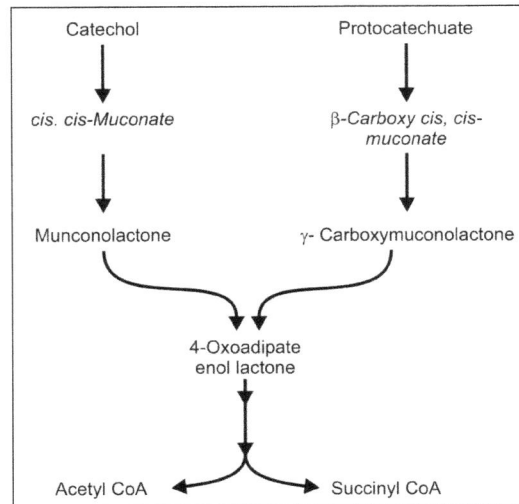

Conversion of Catechol and Protocatechuateto Acetyl CoA and Succinate by ortho-cleavage pathway

Dehalogenation (i.e. removal of a halogen substituent from an organic compound) of halogenated compounds is an essential step for their detoxification. Dehalogenation is frequently catalysed by the enzyme di-oxygenase. In this reaction, there is a replacement of halogen on benzene with a hydroxyl group.

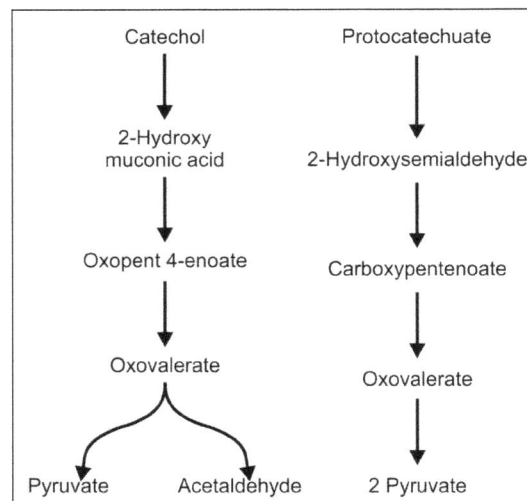

Conversion of Catechol and Protocatechuate to pyruvate and acetalbehyde by meta-cleavage pathway.

Most of the halogenated compounds are also converted to catechol and protocatechuate which can be metabolised. Besides Pseudomonas, other microorganisms such as Azotobacter, Bacilluefs and E. coli are also involved in the microbial degradation of halogenated aromatic compounds.

Biodegradation of Polychlorinated Biphenyls (PCBs)

The aromatic chlorinated compounds possessing biphenyl ring (substituted with chlorine) are the PCBs e.g. pentachlorobiphenyl. PCBs are commercially synthesized, as they are useful for various purposes — as pesticides, in electrical conductivity (in transformers), in paints and adhesives. They are inert, very stable and resistant to corrosion.

However, PCBs have been implicated in cancer, damage to various organs and impaired reproductive function. Their commercial use has been restricted in recent years, and are now used mostly in electrical transformers.

PCBs accumulate in soil sediments due to hydrophobic nature and high bioaccumulation potential. Although they are resistant to biodegradation, some methods have been recently developed for anaerobic and aerobic oxidation by employing a consortium of microorganisms. Pseudomonas, Alkali genes, Corynebacterium and Acinetobacter. For more efficient degradation of PCBs, the microorganisms are grown on biphenyls, so that the enzymes of biodegradation of PCBs are induced.

Biodegradation of some other important compounds:

Organo-nitro Compounds

Some of the toxic organo-nitro compounds can be degraded by microorganisms for their detoxification:

- 2, 4, 6-Trinitrotoluene (TNT): Certain bacterial and fungal species belonging to Pseudomonas and Clostrium can detoxify TNT.

- Nitrocellulose: Hydrolysis, followed by anaerobic nitrification by certain bacteria, degrades nitrocellulose.

- Synthetic detergents: They contain some surfactants (surface active agents) which are not readily biodegradable. Certain bacterial plasmid can degrade surfactants.

Genetic Engineering for more Efficient Bioremediation

Although several microorganisms that can degrade a large number of xenobiotics have been identified, there are many limitations in bioremediation:

- Microbial degradation of organic compounds is a very slow process.

- No single microorganism can degrade all the xenobiotics present in the environmental pollution.

- The growth of the microorganisms may be inhibited by the xenobiotics.

- Certain xenobiotics get adsorbed on to the particulate matter of soil and become unavailable for microbial degradation.

It is never possible to address all the above limitations and carry out an ideal process of bioremediation. Some attempts have been made in recent years to create genetically engineered microorganisms (CEMs) to enhance bioremediation, besides degrading xenobiotics which are highly resistant (recalcitrant) for breakdown. Some of these aspects are briefly described.

Genetic Manipulation by Transfer of Plasmids

The majority of the genes responsible for the synthesis of bio-degradative enzymes are located on the plasmids. It is therefore logical to think of genetic manipulations of plasmids. New strains of bacteria can be created by transfer of plasmids (by conjugation) carrying genes for different degradative pathways.

If the two plasmids contain homologous regions of DNA, recombination occurs between them, resulting in the formation of a larger fused plasmid (with the combined functions of both plasmids). In case of plasmids which do not possess homologous regions of DNA, they can coexist in the bacterium (to which plasmid transfer was done).

The first successful development of a new strain of bacterium (Pseudomonas) by manipulations of plasmid transfer was done by Chakrabarty and his co-workers in 1970s. They used different plasmids and constructed a new bacterium called as superbug that can degrade a number of hydrocarbons of petroleum simultaneously.

United States granted patent to this superbug in 1981 (as per the directive of American Supreme Court). Thus, superbug became the first genetically engineered microorganism to be patented. Superbug has played a significant role in the development of biotechnology industry, although it has not been used for large scale degradation of oil spills.

Creation of Superbug by Transfer of Plasmids

Superbug is a bacterial strain of Pseudomonas that can degrade camphor, octane, xylene and naphthalene. Its creation is depicted.

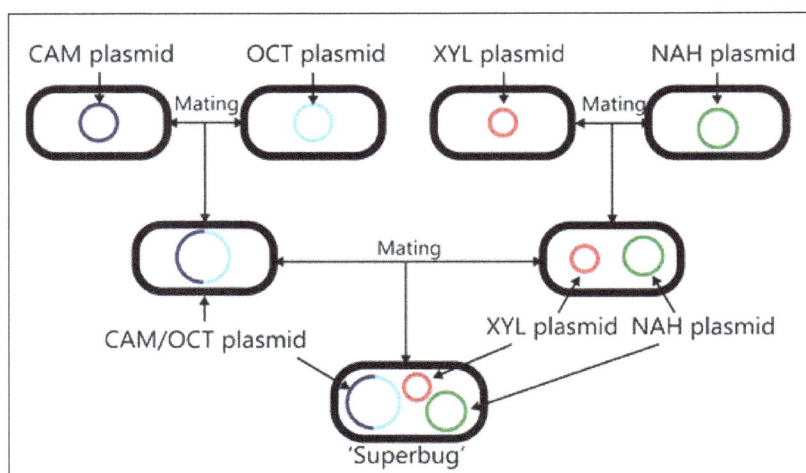

Creation of the Superbug by transfer of plasmids (A.B.C.D.E.F and G are the different starains of bacteria containing the plasmids shown. Strain G is the superbug.)

The bacterium containing CAM (camphor- degrading) plasmid was conjugated with another bacterium with OCT (octane-degrading) plasmid. These plasmids are not compatible and therefore, cannot coexist in the same bacterium. However, due to the presence of homologous regions of DNA, recombination occurs between these two plasmids resulting in a single CAM-OCT plasmid. This new bacterium possesses the degradative genes for both camphor and octane.

Another bacterium with XYL (xylene-degrading) plasmid is conjugated with NAH (naphthalene-degrading) plasmid containing bacterium. XYL and NAH plasmids are compatible and therefore can coexist in the same bacterium. This newly, produced bacterium contains genes for the degradation of xylene and naphthalene.

The next and final step is the conjugation of bacterium containing CAM-OCT plasmid with the other bacterium containing XYL and NAH plasmids. The newly created strain is the superbug that carries CAM-OCT plasmid (to degrade camphor and octane), XYL (xylene-degrading) plasmid and NAH (naphthalene-degrading) plasmid.

Development of Salicylate—Toluene Degrading Bacteria by Plasmid Transfer

Some attempts have been made for the creation of a new strain of the bacterium Pseudomonas putida to simultaneously degrade toluene and salicylate. Toluene-degrading (TOL) plasmid was transferred by conjugation to another bacterium that is capable of degrading salicylate (due to the presence of SAL plasmid).

The newly developed strain of Pseudomonas can simultaneously degrade both toluene and salicylate. And this occurs even at a low temperature (0-5 °C). However, the new bacterium is not in regular use, as more research is being conducted on its merits and demerits.

Genetic Manipulation by Gene Alteration

Work is in progress to manipulate the genes for more efficient biodegradation. The plasmid pWWO of Pseudomonas codes for 12 different enzymes responsible for the meta-cleavage pathway (for the conversion of catechol and protocatechuate to pyruvate and acetaldehyde, for degradation of certain aromatic compounds. Some success has been reported to alter the genes of plasmid pWWO for more efficient degradation of toluene and xylene.

Vermitechnology

Vermitechnology is an important aspect of biotechnology involving the use of earthworms for processing various types of organic waste into valuable resources. It is the latest biotechnology which helps in giving bio-fertilizers in the terms of vermicompost, for agricultural uses and a high quality protein (earthworm biomass) for supplementing the nutritional energy needs of animals, at a faster rate. Vermicompost, specifically earthworm casts, are the final product of vermicomposting. It is an aerobic, bio-oxidation and stabilization of nonhaemophilic process of organic waste decomposition that depends upon earthworms to fragments, mix and promotes microbial activity. Vermicomposting facilities are reported to be already in commercial operation in Japan, Canada, USA and is also being efficiently practiced in Philippines and in Asia. Ghosh reported that vermicomposting at commercial level was started at Ontario (Canada) only in 1970 and is now processing about 75 tones of refuse per week.

Vermicomposting as a principle originates from the fact that earthworms in the process of feeding fragment the substrate thereby increasing its surface area for further microbial colonization. During this process, the important plant nutrients such as nitrogen, potassium, phosphorus and calcium present in the feed material are converted through microbial action into forms that are much more soluble and available to the plants than those in the parent substrate. Earthworms are active feeders on organic waste and while utilizing only a small portion for their body synthesis, they excrete a large part of these consumed waste material in a half digested form. Since the intestine of earthworms harbour wide range of microorganisms, enzymes, hormones, etc., these half digested substrate decomposes rapidly and is transformed into a form of vermicompost with in a short time.

This process takes place in the mesosphilic temperature range (35 – 40 °C). Earthworms prepare organic manures, through their characteristic functions of breaking up organic matter and combines it with soil particles. The final product is a stabilized, well humified, organic fertilizer, with adhesive effects for the soil and stimulator for plant growth and most suitable for agricultural application. The action of earthworms in this process is both physical/mechanical and biochemical. Physical participation in degrading organic substrate results in fragmentation, thereby increasing the surface area of action turnover and aeration. Biochemical changes in the degradation of organic matter are carried out through enzymatic digestion, and of organic and inorganic materials. About 5 – 10 percent of ingested materials is absorbed into the tissue for their growth and metabolic activity and rest is excreted as vermicast. The vermicast is mixed with mucus secretion of the gut wall and of the microbes and transformed into vermicompost. The decomposition process continues even after the release of the cast by the establishment of microorganisms. The studies on the effects of vermicomposting on one components of organic waste showed that vermicompost enhances degree of polymerization of humid substances along with a decrease of ammonium N and an increase of nitrogen.

The key role of earthworms in decomposing the waste materials and improving soil fertility is well known since long. and the term given to the conversion of biodegradable matter by earthworms into vermicast. These droppings are high in nutritive value with many micronutrients like Mn, Fe, Mo, B, Cu, Zn etc. in addition to growth regulators. They are converted into soluble forms, that can be readily available to crop plants. Moreover it is rich in several microflora like Azospirillum, Actinomycetes, Phosphobacillus etc. which multiplies faster through the digestive system of earthworm. It is mixed with the soil and made liquid fertilizer. It influences the physicochemical as well as the biological properties of soil which in turn improve the fertility. Vermicomposting is an innovative, relatively new alternative technology that makes an eco friendly environment

Vermicompost looks peat like material with high porosity, aeration, drainage, water holding capacity and microbial activity. It contains most important nutrients in available forms such as nitrates, phosphates, exchangeable calcium, soluble potassium etc. and has large particulate surface area that provides man sites for microbial activity and for the strong retention of nutrients. The plant growth influencing materials i.e. auxins, cytokinins, humid substances etc.

The nutrient status of vermicompost produced with different organic waste is; organic carbon 9.15 to 17.98 %, total nitrogen 0.5 to 1.5 %, available phosphorus 0.1 to 0.3 % available potassium 0.15, calcium and magnesium 22.70 to 70 mg/100g, copper 2 to 9.3ppm, zinc 5.7 to 11.5ppm and available sulphur, 128 to 548ppm. Several researchers have compared vermicasts with the surrounding soils and reported their results.

Vermiompost is rich in microbial diversity, population, and activity and vermicast contains enzymes such as proteases, amylases, lipase, cellulose and chitinase which continue to disintegrate organic matter even after they have been ejected. The chemical analysis of casts shows 2 times the available magnesium, 5 times the available nitrogen, 7 times the available phosphorus and 11 times the available potassium compared to the surrounding soil. The vermicompost is considered as an excellent product that has reduced the level of contaminants and tends to hold more nutrients over a longer period without impacting the environment.

Being rich in macro and micro-nutrients, the vermicompost, has been found as an ideal organic manure enhancing biomass production of a number of crops. The importance of vermicompost in agriculture, horticulture, waste management and soil conservation has been reviewed by many workers.

The growth of plants with vermicompost, influenced the reduction of bioavailable form of heavy metals and elimination of pathogens. Vermitechnology stimulates the growth of tea and enhance the soil quality which was practiced in association with Parry Agro Ltd., in India involving inoculation of earthworms in trenches with organic inputs in between tea plantations, thereby increasing the production between 75 and 240%. Earthworms ingest large amount of soil and are therefore exposed to heavy metals through their intestine as well as through the skin, therefore concentration of heavy metals from the soil enter their body. Earthworms may serve as bioindicators of soil contaminated with pesticides, and heavy metals. Lead, cadmium, zinc and copper are accumulated and, under some environmental conditions, bioconcentrated in earthworms.

The increased content of K in vermicompost can be attributed largely to gut transit process. The action of earthworm Eisenia fetida with beneficial microorganisms worked upon the substrates and brings the nutrient value more. But due to the combustion of carbon during respiration, the C:N ratio reduced statistically, which is one of the indications of the biodegradation. The earthworm effect on leaf litter indicate the interplay between chemical composition of leaf litter and earthworm activity. Earthworm contributed relatively more to the breakdown of leaves which had a high C:N ratio, lignin and polyphenol The nitrogen content of vermin compost prepared from water hyacinth remains high and suitable for the use of agricultural purpose.

The organic matter needed for vermicomposting can be obtained from aquatic weeds which were freely available in freshwater ponds. Vermicomposts produced with duckweed were characterized by large amounts of dry mass, including organic carbon and ash. However, these vermicomposts still included less ash than the ones produced with manure and household organic wastes, sewage sludge, tannery waste + straw, duckweed + sewage sludge 1:1. In a biological sewage treatment plant, duckweed acts as a natural sewage treating factor by absorbing nutrients, mainly nitrogen and phosphorus. The body fluid and excreta secreted by earthworm (e.g. mucus, high concentration of organic matter, ammonium and urea) promote microbial growth in vermicomposting. Composting of aquatic weed biomass was also used for the purpose of vermicomposting with an aid of Eisenia fetida which is a surface feeding earthworm commonly known as red worm is very active and potentially useful for management of all types of waste. the nutrient value of vermicompost is significantly influenced by the origin, type and proportions of the (mixed) organic wastes utilized.

References

- Environmental-biotechnology, meaning, applications-and-other-details, environmental-biotechnology, bio-technology: biologydiscussion.com, Retrieved 1 May, 2019

- Bioremediation, terms: investopedia.com, Retrieved 4 Feb, 2019

- Environmental-biotechnology: biotechonweb.com, Retrieved 4 January, 2019

- Bioremediation-techniques-used-in-environmental-biotechnology, biotechnology: biologydiscussion.com, Retrieved 13 Febuary, 2019

- Bioremediation, terms: investopedia.com, Retrieved 17 March, 2019

- Biodegradation-and-bioremediation-with-diagram, biodegradation, biotechnology: biologydiscussion.com, Retrieved 1 May, 2019

- Biofuel, technology: britannica.com, Retrieved 25 August, 2019

- Generations-of-biofuels: energyfromwasteandwood.weebly.com, Retrieved 31 July, 2019

- What-biodiesel, biofuels: esru.strath.ac.uk, Retrieved 4 June, 2019

7

Nanobiotechnology and its Applications

Nanobiotechnology is the intersection of nanotechnology and biology. It aims to apply nanotools to relevant medical/biological problems and refine these applications. It is also focused at generating cures and regenerating biological tissues. Science and technology have undergone rapid developments in the past decade that have resulted in the discovery of significant tools and techniques in the field of nanobiotechnology; which have been extensively detailed in this chapter.

Nanobiotechnology is the application of nanotechnologies in biological fields. Chemists, physicists and biologists each view nanotechnology as a branch of their own subject, and collaborations in which they each contribute equally are common. One result is the hybrid field of nanobiotechnology that uses biological starting materials, biological design principles or has biological or medical applications.

While biotechnology deals with metabolic and other physiological processes of biological subjects including microorganisms, in combination with nanotechnology, nanobiotechnology can play a vital role in developing and implementing many useful tools in the study of life.

Although the integration of nanomaterials with biology has led to the development of diagnostic devices, contrast agents, analytical tools, therapy, and drug-delivery vehicles, bionanotechnology research is still in its infancy.

Nanotoxicology

Because of their small size, a large proportion of the atoms that make up a nanoparticle are exposed to the exterior of the particle and would be free to participate in many chemical process. Although the benefits of nanotechnology are widely publicized, discussion of the potential effects of their widespread use in consumer and industrial products is just beginning. Concerns over safety issues are heightened by the fact that the nanotechnology work-force is growing rapidly, projected to reach 2 million workers by 2015. Both pioneers of nanotechnology and its opponents are finding it extremely hard to argue their case because of the limited information available to support one side or the other. In the United Kingdom, the Prince of Wales has requested advice on nanotechnology from the Royal Society, whereas Greenpeace and the Canadian Action Group on Erosion, Technology and Concentration have called for a moratorium on the use of nanoparticles until the toxicological issues have been resolved.

Although some concerns may be ill-founded, it remains true that the toxicology of many nano materials has not yet been fully evaluated. To address this issue, some companies are participating in the European Nanosafe con-sortium, which is starting to evaluate the possible risks presented by nanomaterials. In the United States, the Center for Biological and Environmental Nanotechnology at Rice University has begun an investigation of two pop-ular nanomaterial systems: carbon nanotubes and TiO_2. In vitro diagnostic use does not pose any safety risks to people but there is a concern over nanoparticles use in vivo, particularly those <50 nm in size, which can enter the cells.

There are still many unanswered questions about their fate in the living body. Because the huge diversity of materials used and the wide range in sizes of nanoparticles, these effects will vary a lot. It is conceivable that particular sizes of some materials might turn out to have toxic effects and further investigations will be needed.

The model material responsible for providing neuro-protection is fullereno, which is hydroxyl functionalized fullerene. Yamawaki and Iwai, reported the in vitro toxicity of fullerenols in human umbilical vein endothelialcells that were treated with 1–100 µg/mL concentrations(average diameter 4.7–9.5 nm) for a day, which inducedcytotoxic morphological changes as well as showing cyto-toxicity via lactate dehydrogenase assay in a dose depen-dent manner. An eight-day chronic treatment (10 µg/mL) also inhibited cell attachment and delayed endothelial cellgrowth. Varying biological effects of a single nanomaterialsuch as the hydroxy fullerene offers a clear demonstrationof extraordinary situations where a single nanomaterialplays both beneficial (neuro-protection) and unfavorable(specific cell toxicity response) roles within a biologicalsystem.

Carbon nanotubes, owning to their structural robustnessand synthetic versatility, have been utilized in multiple biomedical applications including tissue engineering. Jan and Kotov, have formulated a nanocompositematrix consisting mainly of single-walled carbon nano-tubes which were utilized as a growth substrate for murineembryonic neural stem cells. Differentiation, growth, and biocompatibility reported by the authors supported positive uses of such nanocomposites. In The FDA approval is essential for clinical applications of nanotechnology, and substantial regulatory problems could be encountered in the approval of nanotechnology-based products. Pharmaceuticals, biological, and devices are all regulated differently by the FDA, and it is not yet clear how emerging nanotherapeutics will be evaluated.

There is concern that, as a consequence of the inter-action of the cells with the nanomaterials, the toxicity of the nanomaterials may also differ from that of the bulk materials, as with the interaction of the cells with the bulk materials. In order to better understand the hazards of materials and develop safer nanomaterials, studies in the nano-bio interface must be made. These studies includeanalysis of the effect of physicochemical properties on cell bioavailability, uptake, and bioprocessing. Studies alsoneed to be made to optimize these properties for their utility in nanomaterials for therapeutic use. Numerous studies on the toxicological effects of nanomaterials are underway. However, a clear understanding of the possible health effects of nanoparticles is still unavailable resulting in limitations to the widespread use of these clearly extraordinary nanobiotechnologies.

Applications of Nanobiotechnology

Medical Applications

Two of the most exciting and promising domains of nanotechnology for advancements are health and medicine. Nanotechnology offers potential developments in pharmaceuticals, medical imaging and diagnosis, cancer treatment, implantable materials, tissue regeneration, and multifunctional platforms combining several of these modes of action.

Diagnosis

One primary goal in nanobiotechnology is the design of new methodologies to diagnose a number of diseases at an early stage with cheaper material and more sophisticated equipment than is possible today. Much research is currently being performed in this area. The utilization of metal and semiconductor nanoparticles in biomedical applications has been demonstrated by many research groups. Aaron et al. Have shown that 25-nmgold nanoparticles when conjugated with anti-epiderma lgrowth factor receptor monoclonal antibodies can be efficiently used as in vivo targeting agents for imaging cancer markers, specifically epidermal growth factor receptors. The Au nanoparticles result in a dramatic increase in sig-nal contrast compared to other antibody-fluorescent dye targeting agents.

Nanobodies have the potential to be a new generation of antibody-based therapeutics and to be used in diagnostics for diseases such as cancer. The advantages of nanobodies to developing therapeutics are the extremely stable and bind antigen with nanomolar affinity, a high target specificity and low toxicity, the ability to combine the advantages of conventional antibodies with important features of small-molecule drugs, and their ability to be produced cost effectively on a large scale. An option for the use of antibodies in molecular biomedical is aptamers. These molecules are chemically stable and easily produce single-stranded nucleic acid molecules. One example of an application of aptamers in diagnosis is the work of Niedzwiecki and colleagues. In this work, nanopores and aptamers were combined to detect a single molecule of the nucleocapsid protein 7 (NCp7), a protein biomarker of theHIV-1 virus, with high sensitivity.

An interesting tool being developed today to be utilized in tumor diagnosis is RNA nanoparticles. Although several researchers are adverse to RNA nanotechnology, due to the susceptibility of RNA to RNase degradationand serum instability, Shu and colleagues have developed a toolkit to obtain stable RNA nanoparticles. In this work, homogeneous RNA nanoparticles were obtained, which targeted cancer exclusively in vivo with-out accumulation in normal organs and tissues. Functionalized nanoparticle aggregating fluorescence imaging techniques, known as quantum dots, have the potentialfor real-time and non-invasive visualization of biological events in vivo. The nanoparticles can provide a solid support for sensing assays with several kinds of ligand molecules attached to each nanoparticle, simplifying assay design. They can also withstand significantly larger number of cycles of excitation and light emissions than typical organic molecules, which more readily decompose, increasing the labeling ratio for higher sensitivity in com-plex biological systems. Another advantage is that these miniaturized fluorescent nanoparticles can also be easily taken up by cells through endocytosis and subsequently used for site-specific intracellular measurements and long-term tracking of biomolecules in real time. This tool is being studied by several groups around the world.

Gene Therapy

Gene therapy is a recently introduced method for the treatment or prevention of genetic disorders by correcting defective genes responsible for disease development based on the delivery of repaired genes or the replacement of incorrect ones. The most common approach for correcting faulty genes is insertion of a normal gene into a nonspecific location within the genome to replace a nonfunctional gene. An abnormal gene could also be swapped for a normal gene through selective reverse mutation, which returns the gene to its normal function. Mammalian cells typically have a diameter of a few microns and their organelles are within the nanometerrange. The use of nanodevices has the advantage of entering the cells more easily when compared to larger devices and they can, therefore, interact better with the cells or at least in a different way.

The use of nanotechnology in gene therapy could be applied to replace the currently used viral vectors with potentially less immunogenic nanosize gene carriers. Delivery, therefore, of repaired genes or the replacement of incorrect genes are fields in which nanoscale objects could be introduced successfully. The use of nanotechnology, here exemplified as the use of nanoparticles, has some advantages in gene delivery: the structure of the nanoparticles protects the nucleic acids from degradation by nucleases and the environment; it also minimizes side effects by directing the nucleic acid to the specific location of action; they facilitate cell entry of nucleic acids and normally nanoparticles sustain gene delivery for longer periods when compared to other vehicles.

Drug Delivery

Controlled delivery systems are used to improve the therapeutic efficacy and safety of drugs by delivering them to the site of action at a rate dictated by the need of the physiological environment, which in turn would reduce both toxicity and side effects. Electrospun nanofibersmay serve as a promising delivery vehicle as a result of their 3D nano-sized features that closely resemble those ofthe ECM. By this technique it is possible to incorporate biological molecules by using an emulsion or directly in apolymer solution.

Co-axial also is used in drug delivery due its capacity of producing micro/nanotubes, drug or protein-embedded nanofibers, and hybrid core–shell nanofibrous materials. Structures built by electro spinning or co-axial permit the liberation of growth factors as epidermal growth fac-tor (EGF), fibroblast growth factor (FGF), transforming growth factor (TGF), bone morphogenetic protein (BMP),and neurotrophins and neurokines, among others, used for neural, endothelial, and bone formation, etc.

Another nanotechnology tool which is under intense investigation for drug delivery is nanoparticles. They can principally be fabricated by lipids and polymers. Polymeric compounds that are currently being used in drug products include poly (DL-lactic-coglycolic acid) (PLGA) polyvinyl alcohol, poly(ethylene-co-vinyl acetate), polyimide, and poly(methy lmethacrylate). Co-delivery is an alternative for the administration of different drugs,which by conventional therapeutic method cannot be used together. Therefore, nanoscale systems can be used to facilitate the delivery of incompatible drugs. They can also be used in the ranostics, in which the particle is used as adevice to diagnose and treat the disease at the same time. These techniques are also used for the liberation of phar-macological agents against several diseases, such as bacte-rial infection, inflammations, and principally cancer,among others. They are also being investigated as a tool for the delivery of drugs through the blood-brain barrier.

Tissue Engineering

 The growing trend of increasing life expectancy of the population as well as the serious limitations in the use of allografts, autologous grafts, or xenografts has led scientists around the world to invest more in the search for alter-natives. Therefore, research in this area aims to apply the principles of cell transplantation and engineering to construct biological substitutes. These, in turn, are used in an attempt to restore and maintain normal function of organs and tissues previously diseased or injured. Thus, the focus of tissue engineering (TE) is to repair or recon-struct lost or damaged tissue through the use of growth factors, cell therapy, injectable biopolymers, and bioma-terials, which serve as support for the development of the cells. Cells interact with the environment around them through thousands of interactions on a nanometric scale. Therefore, the goal of TE on a nanoscale is to createbiomaterials that direct interactions between cells and theirmicro-envi-ronment, by the creation of nanoscale molecu-lar signals of biological interest. Thereby, the cells receive, process, and respond to information presented in the sur-rounding environment, these actions being essential for thecontrol of cell behavior.

From the techniques used to construct biomaterials to be cultivated with cells, electrospinning is the most widely studied and it has also been demonstrated to give the most promising results in terms of TE applications. It is a highly versatile method of transforming solutions, mainly made from polymers, in continuous filaments with diameters ranging from a few micrometers to nano-meters. Through this method, the fibers can be obtained randomly or in an ordered way. Electro-spinning works by the electrostatic principle, where the solution is supplied to the system viaa syringe and is subjected to a difference in electrical voltage, yielding solid fiber at the end of the process. Co-axial electro spinning is a modification or extension of the electrospinning technique. Using this compound spinneret method, two components can be fed through different co-axial capillary channels and are integrated into a core—shell structured composite fiber. The nanofibers obtained by both methods have been used in many biological applications, as follows. The skin is an organ that serves as a physical barrier to the external environment and consists primarily of epidermis and dermis. Because the skin acts as a protective barrier, any injury caused to it should be repaired quickly and efficiently.

Especially in large burns and chronic wounds, the treatments available are insufficient to prevent scar for mation and promote healing of the patient. Thus, the regeneration of skin is an import-ant field for TE. The substitutes fabricated for use in TE normally act as supplementary der-mal templates and improve wound healing. Several stud-ies are being conducted in an attempt to cre-ate an ideal substitute that could eliminate the limitations encountered so far in the current skin substitutes. Most of them use polymeric matrices associated with stem cells and, in some cases, biological agents, such as growth factors and plant substances are also used. Other cells, such as fibroblasts and keratinocytes are largely used too. In the central nervous system, degeneration of neurons or glial cells or any unfavorable change in the extra-cellular matrix of neural tissue can lead to a wide variety of clinical disorders, such as Alzheimer's, Huntington's, and Park in son's diseases, as well as traumatic damage such as spinal cord in jury. The biggest concern here is that almost all diseases in this system lead to irreversible loss of functions. How-ever, research in this field with the electro spinning technique is still not well explored by bioengineers.

nerve gaps in the peripheral nervous system. For this kind of regeneration, the orientation of the fibers is one of the parameters most studied when considering only the morphology of the bio-materials. Subramanian et al. compared random and longitudinally aligned scaffolds of poly(lac-tic-co-glycolic acid) (PLGA) for the cultivation of Schwann cells. The results for morphological and cell pro-liferation assays demonstrated that the aligned fibers assist in the direction of Schwann cells and have a better pro-life ration rate than random fibers. The physic-chemical parameters evaluated also showed better results for nerveregeneration than random nanofibers.

Bone injuries occur for reasons including degenerative, surgical, and traumatic processes, resulting in severe pain and disability for millions of people worldwide. The design of scaffolds for bone TE is based on the physical properties of bone tissue, such as mechanical strength, pore size, porosity, hardness, and overall 3D architecture. Currently, the most widely used synthetic bioactive bone substitute is calcium phosphate-basedmaterials. Hydroxyapatite (HAp) is chemically similar to the inorganic component of bone matrix, and it is osteo-conducting as well as osteo-inducting. Several studies have been performed with the development of nanofibers incorporated with HAp, showing good results.

Because of its limited capacity for spontaneous repair due to its lack of vascularity and poor vailability of chondrocytes and progenitor cells, cartilage cannot be restored to its normal function and structure after damage caused by trauma, disease, and accidents. Therefore, TE as a potential approach to regenerate cartilage tissue holds good promise. The cells that could be used to treat iseases of cartilage can be either differentiated cells, such as chondrocytes, or undifferentiated cells, such as mesenchymal stem cells. To date, different scaffolds have been used to repair cartilage defects; in a few studies, the scaffolds have been incorporated with appropriate cells and their capacity for in vivo healing has been compared with cell-free scaffolds. In most of the animals, cartilage regeneration was accelerated and improved when using cell-scaffold complexes.

Various injuries and diseases, such as cancer, trauma, infection, inflammation, and congenital abnormalities may result in compromised bladder structure and compliance, requiring reconstruction. In order to effectively replace the diseased bladders of patients, it is necessary to design a bladder tissue replacement construction with increased efficacy. In this way, reabsorbable polymers have been preferable because polymers that do not degrade carry the permanent risk of infection, calcification and unfavourable connective tissue response. Following this reasoning, some studies are being performed, utilizing the association of biodegradable and biocompatible polymers with cells. Thapa et al. in a study using PLGA and poly(ether urethane) (PU) showed that the topography of the scaffold influences cell response. In the results they observed that material that mimics the nanometer topography of native bladder tissue is preferred by the cells, leading to better tissue integration in vivo.

Another interesting observation of Baker et al. when studying the use of polystyrene to create electrospun scaffolds with varying spatial configurations was that scaffolds with longitudinally aligned fibers demonstrate similar strength to native bladder tissue and greater strength than transversely aligned and randomly aligned scaffolds and also that cell alignment was greater when grown on scaffolds with aligned fibers. One in five people will develop heart failure in their life time. Such a high risk is fuelled by the intrinsic inability of the heart to regenerate itself after injury. Because of this high incidence, the main goal of bioengineers in this field is to engineer tissues that are capable of establishing normal heart contractile function and prevent pathological remodeling. The alignment of cardiomyocytes is important for the contraction and impulse propagation along the long axis of the cells. Aligning cardiomyocytes in 3D provides adequate functionality to replace damaged tissues, and the electrospinning technique has been shown to be efficient in creating scaffolds that supply this necessity. Structures, such as core–shell, associated with cells are also applied in the treatment of myocardial infarction, increasing cell transplant retention and survival within the infarct, compared to the standard cell injection system.

Pathogen Detection

There have been recent outbreaks of food disease, most notably the outbreak of Escherichia coli in Germany in which a total of 3,602 cases were reported and 47 people died. Clearly there is a need to monitor food-borne pathogens throughout the food chain from production, processing, and distribution to the point-of-sale. Pathogens may be present in low numbers in a sample for analysis, making detection difficult. Traditional detection methods for pathogen determination like colony count estimation can be laborious and time consuming with completion ranging from 24 h for E. coli to 7 days for Listeria monocytogenes, and these pose significant difficulties for quality control of semi-perishable foods. Pathogen numbers can also be underestimated using these methods due to microorganisms entering viable but non-culturable states due to environmental stress. Upon resuscitation from this state by, for example, an increase in temperature of the cells, microorganisms can regain the ability to cause infection, thus posing a health risk.

Advances in the manipulation of these nanomaterials permit binding of different biomolecules such as bacteria, toxins, proteins, and nucleic acids. One of the major advantages of using nanomaterials for biosensing is that because of their large surface area, a greater number of biomolecules are allowed to be immobilized and this consequently increases the number of reaction sites available for interaction with a target species. This property, coupled with excellent electronic and optical properties, facilitate the use of nanomaterials in label-free detection and in the development of biosensors with enhanced sensitivities and improved response times. Biosensors are currently used in the areas of target identification, validation, assay development, lead optimization and absorption, distribution, metabolism, excretion, and toxicity. They are best suited to applications using soluble molecules and overcome many of the limitations that arise with cell-based assays. Biosensors are particular useful in the study of receptors because they do not require the receptor to be removed from the lipid membrane of the cell, which can be necessary with other assay methods.

Single-walled carbon nanotubes have been used as a platform for investigating surface–protein and protein– protein binding, as well as to develop highly specific electronic biomolecule detectors. Non-specific binding on nanotubes, a phenomenon found with a wide range of proteins, is overcome by the immobilization of polyethyleneoxide chains. A general method is followed that entails the selective recognition and binding of target proteins, conjugating their specific receptors to polyethylene-oxide functionalized nanotubes. These arrays are attractive because no labeling is required, and the entire assay can be done in the solution phase. This combined with the sensitivity of nanotube electronic devices, provides highly specific electronic sensors for detecting clinically important biomolecules, such as a biomolecules, such as antibodies associated with human autoimmune diseases.

Quantum dots (QDs) are colloidal semiconducting fluorescent nanoparticles consisting of a semiconductor material core (cadmium mixed with selenium or tellurium), which has been coated with an additional semiconductor shell (usually zinc sulfide). Due to their unique size dependent fluorescence properties and photostability, QDs are widely used in place of traditional fluorescent dyes (Fluorescein isothiocyanate FITC). Functionalized QDs have been used as labels for DNA probing of genomic DNA and in fluorescent in situ hybridization (FISH) assays. Metallic nanoparticles such as gold and silver have been used for signal amplification in numerous biodiagnostic devices. Gold nanoparticles have been used in a variety of optical and electrical assays. The electrical properties of the gold nanoparticles were harnessed for the development of a piezoelectric biosensor, for real-time detection of a food-borne pathogen.

Elemental silver and silver salts have been well known as antimicrobial agents in curative and preventive health care for centuries. The antimicrobial activity of the silver salts and complexes (ionic silver) is generally based on the bonding of metallic ions in various biomacromolecular components. Cationic silver targets and binds to negatively charged components of proteins and nucleic acids, thereby causing structural changes and deformations in bacterial cell walls, membranes, and nucleic acids. Silver ions are generally well known to interact with a number of electron donor functional groups like thiols, phosphates, hydroxyls, imidazoles, indoles, and amines. Accordingly, it is believed that silver ions that bind to DNA block transcription while those that bind to cell surface components interrupt bacterial respiration and adenosine triphosphate (ATP) synthesis. Other reports suggest that silver ions block the respiratory chain of microorganisms in the cytochrome oxidase and nicotinamide adenine dinucleotide (NADH) succinate dehydrogenase region, the latter being an enzyme complex in bacterial cells.

Food Safety

The research in the area of nanobiotechnology in food involves mainly adding antioxidants, antimicrobial, biosensors, and other nanomaterials at packaging. The medical, pharmaceutical, and cosmetics industries have been using nanoparticles made from food to improve the characteristics of its products. Nanobiotechnology in food packaging has been a focus in recent years. The potential perspectives of bio-nanocomposites for food packaging applications together with bio-based materials, such as edible and biodegradable nanocomposite films have gained attention. Among the available metal nanoparticles, silver and related materials have been utilized in many nano-based commercial products for their antimicrobial property. Studies suggest that the antimicrobial performance is enhanced due to an intensive surface area/reduced particle size.

The use of nanomaterials as antimicrobial agents has received increased attention in recent years. Since the antimicrobial properties of silver are well documented it is no surprise that the antimicrobial activity of AgNPs dominate publications in this area. Others nanoparticles as zinc and sulfur, showed antimicrobial activity according to some authors. The combination of silver nanoparticles with water soluble biopolymers will produce new antimicrobials. Based on this, various natural polymers such as gum acacia, starch, gelatin, sodium alginate, and carboxy methyl cellulose have been employed to prepare biocompatible polymeric silver nanocomposites. Chitosan, a natural polymer composed of poly(-(1- 4)-2-amino-2-deoxy-D-glucose), is one of the structural polysaccharides that is abundantly available in nature after cellulose. Chitosan interacts very easily with bacterium and binds to DNA, glycosaminoglycans, and most of the proteins, thereby enhancing the antimicrobial effect of silver nanoparticles.

The developed silver nanocomposite films have exhibited fairly good mechanical strength and superior antimicrobial properties. Further, the current work demonstrates a promising method to combine silver nanocomposites with a natural compound (curcumin) in developing novel antimicrobial agents. These agents may find potential applications in antimicrobial packaging materials and wound dressing/wound burns. Vivekanandhan et al. impregnated AgNPs into microcrystalline cellulose using Murraya koenigii extract as the reducing agent at ambient conditions and extending their application into the fabrication of polylactic acid (PLA) based antimicrobial bionanocomposite films. Costa et al. evaluated the effects of silver-montmorillonite nanoparticles (Ag-MMT) on microbial and sensorial quality decay during storage of a packaged fresh fruit salad. The authors observed that the shelf life increased about 5 days when using the sample with 20 mg of Ag-MMT. To sum up, it can be concluded that Ag-MMT nanoparticles optimized in a proper packaging system could represent a solution to control the quality decay of fresh-cut fruit.

Sastry et al. identified areas of nanoresearch determinants of food security, which are productivity, soil health, water security and storage, and distribution of food. For the distribution of food, the thematic areas are food processing and food packaging. For these, the nanoresearch areas are nanoscale phenomena, nanoparticles, biosensors, nanofibers, among others. Some applications for these areas are natural biopolymer-based nanocomposite films used for food packaging for safe storage, nanowire immunosensors array for the detection of microbial pathogens, quick detection of food-borne pathogens using bioconjugated nanomaterials, biosensor, nanocantilevers and carbon nanotubes, and nanoscale titanium dioxide particles as a blocking agent of UV light in plastic packaging. Gonçalves et al. used curcumin that is extracted from the rhizomes of the Curcuma species and is a natural polyphenol with antioxidative, anti-inflammatory, and anti-cancer properties to produce a nanogel. The authors reached a dextrin nanogel that served as an effective "nanocarrier" for the formulation of lipophilic curcumin by increasing its water solubility, improving its stability, and controlling its release profile. The small size of this system can be advantageous for passive targeting of tumor tissues. Butnariu and Giuchici, studied the therapeutic effects of nanoemulsions made of an aqueous extract of propolis and lycopene. The results showed that inflammation decreased up to 100% using nanoemulsions and a lycopene antioxidant as a nanoemulsion component. In addition, its moisturizer characteristic improves the ability of the skin to defend against sunlight.

Biosurfactants

Biosurfactants are surface active substances that reduce interfacial tension and are produced or excreted at the microbial cell surface. Biosurfactants have been tested in environmental applications, cosmetics, foods, and pharmaceutical industries but also as industrial cleaners and chemical products for agricultural use. Rhamnolipid, a biosurfactant, has been used as the capping agent. Rhamnolipid derived from Pseudomonas aeruginosa has a simple molecular structure, low molecular weight, and high affinity for metal ions. The properties of nanoparticles are controlled by a set of processes called "capping." Therefore the properties of capping agents play an important role in the synthesis of nanoparticles. Narayanan et al. studied the synthesis of rhamnolipid-capped ZnS nanospheres in an aqueous environment without any organic solvent. The capped nanoparticles were stable and water soluble. Rhamnolipid-capped stable metal nanoparticles could be used to make electronic and optoelectronic devices.

Surfactin is a type of lipopeptide produced by Bacillus subtilis. It is considered one of the most powerful biosurfactants, as it is able to reduce the surface tension from 72 to 27 mN m−1 at a concentration as low as 0.005%. Cadmium sulfide nanoparticles (CdS-NPs) are typical II–VI semiconductors with unique optical properties and tunable photoluminescence. They have potential applications in solar energy conversion, nonlinear optical, photoelectrochemical cells, among others. Surfactin can be used as a biodegradable and biologically compatible stabilizing agent in the synthesis of stable CdS-NPs. The simple and inexpensive procedure of obtaining surfactin offers an advantage of its application in the production of CdS-NPs, which remains stable up to six months without compromising their functionality. Thus, this environmentally friendly process has significance in nanobiotechnology for the large-scale production of highly stable metal nanoparticles and QDs. Through many biological and medical studies, it has been found that artificially made calcium phosphate has characteristics such as biocompatibility and bioactivity, which arise from alteration of its size, morphology, stoichiometry, or the composition of calcium phosphate.

The main components of the microemulsion process are a mixture of oil, water, and surfactant. The surfactant is used to create a low interfacial tension, which aids in the production of nanosized particles. According to previous research cited by Maity et al. This technique is feasible for industrial applications to produce nanoparticles. As a result, the reverse microemulsion process should be suitable for synthesizing nanoscale brushite calcium phosphates with different morphological structures that contain a water/oil interface in the presence of surfactin. According to Maity et al. it has been shown that microemulsions created by surfactin successfully produced calcium phosphate of unique shapes and sizes. The reverse microemulsion process resulted in the formation and growth of brushite nanoparticles under noncalcinated conditions at room temperature. The crystalline structure of the calcium phosphate depended on the ratio of the calcium nitrate tetrahydrate and ammonium phosphate water solution to surfactin. The different structures obtained included hexagonal-shaped, thin-layered, needleshaped, and small roundish-shaped crystals. All the particles produced were nanosized (16–200 nm). Liu et al. studied the surfactin effect on the aggregate size of the PC liposome system, which was carried out by dynamic light scattering measurement. Surfactin was originally obtained from the cell-free broth of Bacillus subtilis HSO121. For pure PC liposome, there is one peak around 120 nm. When the surfactin concentration reaches 0.24 mg/ml, the mixture of PC/surfactin aggregates take on a sheet undulation character and some have a spherical structure with 10–15 nm small diameters, supporting the assumption that these small aggregations are PC/surfactin mixed micelles.

Biosurfactants with self-assembling properties are of particular interest to nanotechnological applications. Rhamnolipids also self-assemble into a variety of structures, and their assembled structures drastically change by slight variation in the headgroup. Because of their carboxylic acid on the headgroups, rhamnolipids reversibly change their self-assembled structures depending on the solution pH. The rhamnolipids form micelles at a pH of more than 6.8, lipid particles at pH 6.6–6.2, lamellar structures at pH 6.5–6.0, and finally vesicles of diameter 50–100 nm at pH 5.8–4.3. The producers of rhamnolipids, which are mostly hydrocarbon-assimilating bacteria, grow well under a narrow pH range around 7.0, but hardly grow under acidic conditions. There have been a lot of natural and synthetic amphiphiles that are able to form selfassembled nanostructures. Sophorolipids form nm-size micelles in water and depending on their concentration, various geometries can be obtained. This property allows their use as a structure-directing agent in the synthesis of nanostructured silica thin films. These nanomaterials have a high specific surface area and tunable pore size distribution, rendering them ideal for various applications such as catalysis, filtration, sensing, and photovoltaic electrodes.

Industrial Biotechnology

The application of biotechnology for industrial purposes is known as industrial biotechnology. It makes use of microorganisms and enzymes for the production of goods such as plastics, food and chemicals. The topics elaborated in this chapter will help in gaining a better perspective about the branches and applications of industrial biotechnology.

Industrial biotechnology is one of the most promising new approaches to pollution prevention, resource conservation, and cost reduction. It is often referred to as the third wave in biotechnology. If developed to its full potential, industrial biotechnology may have a larger impact on the world than health care and agricultural biotechnology. It offers businesses a way to reduce costs and create new markets while protecting the environment. Also, since many of its products do not require the lengthy review times that drug products must undergo, it's a quicker, easier pathway to the market. Today, new industrial processes can be taken from lab study to commercial application in two to five years, compared to up to a decade for drugs.

The application of biotechnology to industrial processes is not only transforming how we manufacture products but is also providing us with new products that could not even be imagined a few years ago. Because industrial biotechnology is so new, its benefits are still not well known or understood by industry, policymakers, or consumers.

From the beginning, industrial biotechnology has integrated product improvements with pollution prevention. Nothing illustrates this better than the way industrial biotechnology solved the phosphate water pollution problems in the 1970s caused by the use of phosphates in laundry detergent. Biotechnology companies developed enzymes that removed stains from clothing better than phosphates, thus enabling replacement of a polluting material with a non-polluting biobased additive while improving the performance of the end product. This innovation dramatically reduced phosphate-related algal blooms in surface waters around the globe, and simultaneously enabled consumers to get their clothes cleaner with lower wash water temperatures and concomitant energy savings.

Rudimentary industrial biotechnology actually dates back to at least 6000 B.C. when Neolithic cultures fermented grapes to make wine, and Babylonians used microbial yeasts to make beer. Over time, mankind's knowledge of fermentation increased, enabling the production of cheese, yogurt, vinegar, and other food products. In the 1800s, Louis Pasteur proved that fermentation was the result of microbial activity. Then in 1928, Sir Alexander Fleming extracted penicillin from mold. In the 1940s, large-scale fermentation techniques were developed to make industrial quantities of this wonder drug. Not until after World War II, however, did the biotechnology revolution begin, giving rise to modern industrial biotechnology.

Since that time, industrial biotechnology has produced enzymes for use in our daily lives and for the manufacturing sector. For instance, meat tenderizer is an enzyme and some contact lens cleaning fluids contain enzymes to remove sticky protein deposits. In the main, industrial biotechnology involves the microbial production of enzymes, which are specialized proteins. These enzymes have evolved in nature to be super-performing biocatalysts that facilitate and speed-up complex biochemical reactions. These amazing enzyme catalysts are what make industrial biotechnology such a powerful new technology.

Industrial biotechnology involves working with nature to maximize and optimize existing biochemical pathways that can be used in manufacturing. The industrial biotechnology revolution rides on a series of related developments in three fields of study of detailed information derived from the cell: genomics, proteomics, and bioinformatics. As a result, scientists can apply new techniques to a large number of microorganisms ranging from bacteria, yeasts, and fungi to marine diatoms and protozoa.

Industrial biotechnology companies use many specialized techniques to find and improve nature's enzymes. Information from genomic studies on microorganisms is helping researchers capitalize on the wealth of genetic diversity in microbial populations. Researchers first search for enzyme-producing microorganisms in the natural environment and then use DNA probes to search at the molecular level for genes that produce enzymes with specific biocatalytic capabilities. Once isolated, such enzymes can be identified and characterized for their ability to function in specific industrial processes. If necessary, they can be improved with biotechnology techniques.

Many biocatalytic tools are rapidly becoming available for industrial applications because of the recent and dramatic advances in biotechnology techniques. In many cases, the biocatalysts or whole-cell processes are so new that many chemical engineers and product development specialists in the private sector are not yet aware that they are available for deployment. This is a good example of a "technology gap" where there is a lag between availability and widespread use of a new technology. This gap must be overcome to accelerate progress in developing more economic and sustainable manufacturing processes through the integration of biotechnology.

Applications of Industrial Biotechnology

The various applications of industrial biotechnology are:

1. Improvement in Fermentation Products:

This achievement can be done in different ways – by selection of improved strain, by transgene application into the microorganism, by using cheaper raw material, by manipulation of medium constituent as well as by simulation of the reactor (adjustment of different cultural conditions like pH, temp., etc.)

Products of microbial fermentation include primary metabolites, secondary metabolites, enzymes, proteins, capsular polysaccharides and cellular biomass (single cell protein).

2. Microbial Production of Synthetic Fuels:

Important fuels can be produced by using many microbes which include ethanol, methane, hydrogen and hydrocarbons. Zymomonas mobilis produces ethanol twice as rapidly as yeasts from carbohydrates. Methane which is used in various industrial purposes can be produced by Clostridia, Bacteriodes, Sclenomonas, Butyrovibrio, etc. from the waste.

3. Microbial Mining or Bioleaching:

The process of bioleaching recovers metals from ores which are not suitable for direct smelting because of their low content. The application of bioleaching process is of particular interest in case of uranium ore. Thiobacillus ferroxidans is the commonest organism which is involved in case of copper and uranium ore processing.

4. Microbial Biomass and Single Cell Protein Production:

Microbial product of commercial significance is the microbial biomass (the microbial cells themselves), e.g., commercially produced yeast cells, bacteria (Methylophilus methylotrophus), flavoring cheese from fungal biomass (Penicillium roquefortii).

Single cell proteins are the dried cells of microorganisms such as algae, certain bacteria, yeasts, moulds and some higher fungi. The protein percentages for various single cell proteins are high.

5. Production of Enzymes and Human Proteins:

Bulk of the enzymes is obtained from microbial source by fermentation process. Now several plant enzymes are being used like 'papain' from Carica papaya, 'bromelain' from Ananas cosmosus, 'ficin' from Ficus glabrata. These enzymes are used in meat tenderizing, as protein hydro lysates in beer industry, in clinical application, etc.

Recombinant DNA technology can be used for the improved production of enzymes and proteins. There are many human proteins which have long been believed or known to have therapeutic potential and their increased production has been achieved by using recombinant DNA technology.

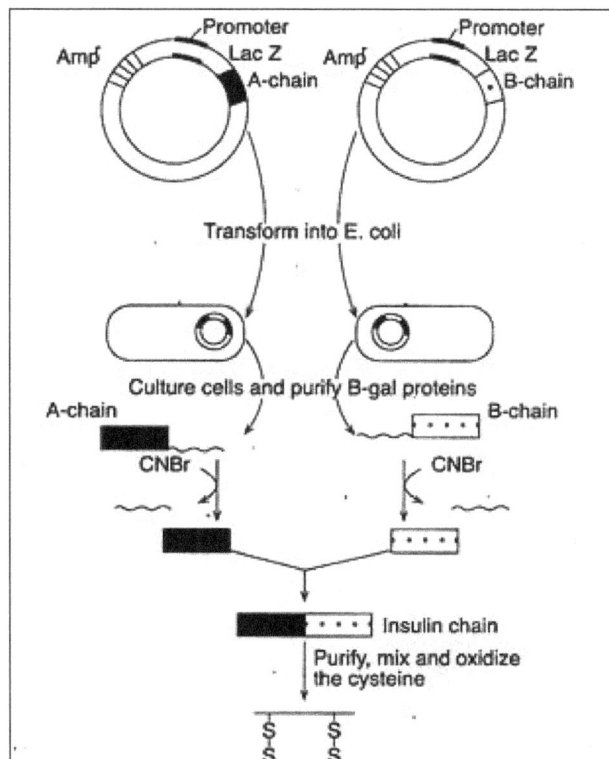

Production of recombinant insulin in E. Coli.

The gene coding for insulin with two polypeptides has been synthesized. Each synthetic gene was linked to a plasmid near the end of the β-galactosidase gene of E. coli. After the gene expression and the translation of mRNA into protein, the two polypeptides were cleaved from the enzyme and linked to form the complete insulin molecule.

Different workers also synthesized complementary DNA from RNA of rat pancreas with the help of reverse transcriptase which was inserted into pBR 322 plasmid in the middle of the gene for penicillinase. The plasmid also contained the structural genes for pro-insulin. The hybrid protein synthesized in the bacterial cell was penicillinase + pro-insulin from which insulin could be separated by trypsin.

Production of interferon took a vital position when human leucocyte interferon was engineered by yeast cells. A DNA sequence coding for human leucocyte interferon was attached to the yeast, alcohol dehydrogenase gene in a plasmid and introduced into ceils of Saccharomyces cerevisae.

The first human peptide hormone synthesized in a bacterial cell was somatostatin, which is one of a group of hormones secreted by hypothalamus, controls the release of several hormones from the pituitary.

The synthetic gene has been inserted into the plasmid, expression vector was constructed from the plasmid pBR 322, to which was added the control region and most of the β-galactosidase gene from the bacterial lac operon; and the gene was inserted next to β-galactosidase.

After the plasmid was introduced into the cells of the bacterium E. coli, the hormone was synthesized as a short peptide tail at the end of the enzyme.

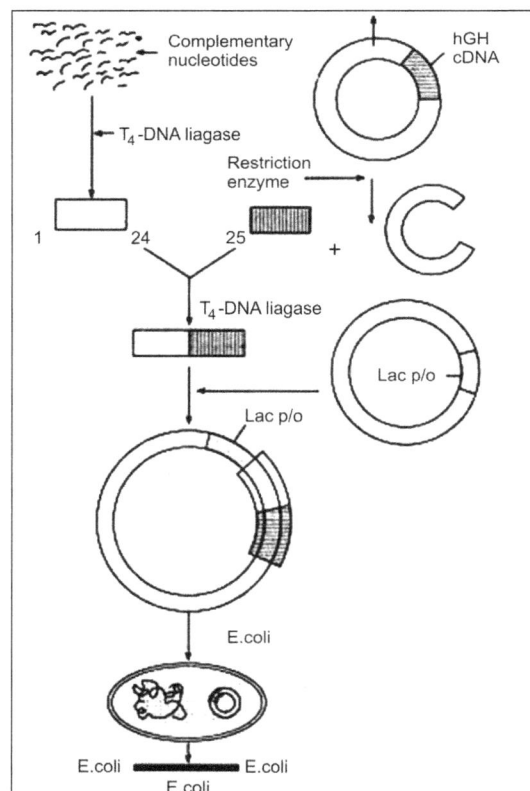

Pression of human growth hormone (hGH) in E. coli, Lac p/o – lactose promoter/operator.

With the advent of techniques of recombinant DNA and gene cloning, several other human hormones are being produced on a commercial scale by isolating specific DNA sequences coding for those proteins/hormones. This is likely to enable clinical application and improve economic provision for their utility in several deficiencies.

6. Production of Secondary Metabolites from Cultured Plant Cell:

In recent years it has been shown that spectrum of compounds can be produced in culture which is beyond the ability of whole plants. By using different precursors several novel compounds of biomedical importance can be obtained.

Pharmaceutical compounds like shikonin is being produced as secondary products with the use of two-stage bioreactor by stimulating the growth phase with the application of different growth regulators.

Serpentine can be obtained from Catharanthus, pseudoephedrine from Ephedra. Plant cells also can be used to accomplish certain changes in the structure and composition of some industrially important chemicals. This conversion by means of a biological system is termed as biotransformation, e.g., digoxin, a cardiovascular drug, produced from digitoxin obtained from Digitalis lanata.

7. Molecular Farming for Healthcare Products:

Transgenic plants Pan be used as 'factories' for production of speciality chemicals and pharmaceuticals like sugars, fatty acids, wax materials as well as antibodies, edible vaccines. The progress is so far reaching that human antibody production through plant seeds has been achieved.

The method involves the introduction of heavy and light chains of immunoglobin genes into microbial vectors. In the next step, these are introduced into the leaf cells of two plants and cultured in vitro for regeneration of the plants.

The plants – one containing heavy and the other with light chain, are then hybridized. The hybrid brings light and heavy chains together, to form the complete immunoglobin (IgA + IgB) in the seeds. This hybrid plant can be utilized for large scale production of seeds containing antibody proteins.

Even antibodies presumed to be effective against cancer have been secured in tobacco seeds. The method is thus a synthesis of recombinant DNA, in vitro technique and conventional hybridization.

Hepatitis B surface antigen is produced in tobacco, rabies virus glycoprotein is produced in tomato, cholera toxin (S-subunit is being produced in potato and tobacco. Transgenic plants are being used as a source of antibodies which provide passive immunization.

One of the recent discoveries in the area of plant biotechnology, is the development of oral vaccine utilizing plant systems. The principle involves, the development of transgenic plants, containing subunits of toxic virus sequences or enterotoxin genes of bacteria like E. coli or Vibrio cholera.

The oral administration of potato or tobacco transgenic tissues led to the development of immunoglobin G and A antibodies. As such, oral administration of plant tissues for production of antibodies in the system is, in effect, a vaccination – the application of vaccine being oral. Such recombinant vaccines, may prove to be a cheaper substitute for expensive vaccination, both in terms of production and administration.

Major Products of Industrial Biotechnology

There are many products which are produced There are many products which are produced in the industries by the help of fermentation technology.

Antibiotics

Many antibiotics are produced by the help of microorganisms such as actinomycetes (e.g., Streptomyces) and filamentous fungi. Following are some antibiotics which are produced by the help of fermentation technology.

Raw Materials

The compounds that make the fermentation broth are the primary raw materials required for antibiotic production. This broth is an aqueous solution made up of all of the ingredients necessary for the proliferation of the microorganisms.

Typically, it contains a carbon source like molasses, or soy meal, both of which are made up of lactose and glucose sugars. These materials are needed as a food source for the organisms. Nitrogen is another necessary compound in the metabolic cycles of the organisms. For this reason, an ammonia salt is typically used.

Additionally, trace elements needed for the proper growth of the antibiotic-producing organisms are included. These are components such as phosphorus, sulfur, magnesium, zinc, iron, and copper introduced through water soluble salts. To prevent foaming during fermentation, anti-foaming agents such as lard oil, octadecanol, and silicones are used.

Manufacturing Process

Although most antibiotics occur in nature, they are not normally available in the quantities necessary for large-scale production. For this reason, a fermentation process was developed. It involves isolating a desired microorganism, fuelling growth of the culture and refining and isolating the final antibiotic product. It is important that sterile conditions be maintained throughout the manufacturing process, because contamination by foreign microbes will ruin the fermentation.

Starting the Culture

1. Before fermentation can begin, the desired antibiotic-producing organism must be isolated and its numbers must be increased by many times. To do this, a starter culture from a sample of previously isolated, cold-stored organisms is created in the lab. In order to grow the initial culture, a sample of the organism is transferred to an agar-containing plate.

    The initial culture is then put into shake-flasks along with food and other nutrients necessary for growth. This creates a suspension, which can be transferred to seed tanks for further growth.

2. The seed tanks are steel tanks designed to provide an ideal environment for growing microorganisms. They are filled with all the things the specific microorganism would need to survive and thrive, including warm water and carbohydrate foods like lactose or glucose sugars. Additionally, they contain other necessary carbon sources, such as acetic acid, alcohols, or hydrocarbons, and nitrogen sources like ammonia salts.

Growth factors like vitamins, amino acids, and minor nutrients round out the composition of the seed tank contents. The seed tanks are equipped with mixers, which keep the growth medium moving, and a pump to deliver sterilized, filtered air. After about 24-28 hours, the material in the seed tanks is transferred to the primary fermentation tanks.

Fermentation

The fermentation tank is essentially a larger version of the steel seed tank, which is able to hold about 30,000 gallons. It is filled with the same growth media found in the seed tank and also provides an environment inductive to growth. Here the microorganisms are allowed to grow and multiply. During this process, they excrete large quantities of the desired antibiotic.

The tanks are cooled to keep the temperature between 73-81 °F (23-27.2 °C). It is constantly agitated, and a continuous stream of sterilized air is pumped into it. For this reason, anti-foaming agents are periodically added. Since pH control is vital for optimal growth, acids or bases are added to the tank as necessary.

Isolation and Purification

After three to five days, the maximum amount of antibiotic will have been produced and the isolation process can begin. Depending on the specific antibiotic produced, the fermentation broth is processed by various purification methods. For example, for antibiotic compounds that are water soluble, an ion-exchange method may be used for purification.

In this method, the compound is first separated from the waste organic materials in the broth and then sent through equipment, which separates the other water-soluble compounds from the desired one. To isolate an oil-soluble antibiotic such as penicillin, a solvent extraction method is used.

In this method, the broth is treated with organic solvents such as butyl acetate or methyl isobutyl ketone, which can specifically dissolve the antibiotic. The dissolved antibiotic is then recovered using various organic chemical means. At the end of this step, the manufacturer is typically left with a purified powdered form of the antibiotic, which can be further refined into different product types.

Refining

1. Antibiotic products can take on many different forms. They can be sold in solutions for intravenous bags or syringes, in pill or gel capsule form, or they may be sold as powders, which are incorporated into topical ointments. Depending on the final form of the antibiotic, various refining steps may be taken after the initial isolation.

    For intravenous bags, the crystalline antibiotic can be dissolved in a solution, put in the bag, which is then hermetically sealed. For gel capsules, the powdered antibiotic is physically filled into the bottom half of a capsule, then the top half is mechanically put in place. When used in topical ointments, the antibiotic is mixed into the ointment.

2. From this point, the antibiotic product is transported to the final packaging stations. Here, the products are stacked and put in boxes. They are loaded up on trucks and transported to various distributors, hospitals, and pharmacies. The entire process of fermentation, recovery, and processing can take anywhere from five to eight days.

Quality Control

Quality control is of utmost importance in the production of antibiotics. Since it involves a fermentation process, steps must be taken to ensure that absolutely no contamination is introduced at any point during production. To this end, the medium and all of the processing equipment are thoroughly steam sterilized.

During manufacturing, the quality of all the compounds is checked on a regular basis. Of particular importance are frequent checks of the condition of the microorganism culture during fermentation. These are accomplished using various chromatography techniques. Also, various physical and chemical properties of the finished product are checked such as pH, melting point, and moisture content.

In the United States, antibiotic production is highly regulated by the Food and Drug Administration (FDA). Depending on the application and type of antibiotic, more or less testing must be completed. For example, the FDA requires that for certain antibiotics each batch must be checked by them for effectiveness and purity. Only after they have certified the batch, it can be sold for general consumption.

The Future

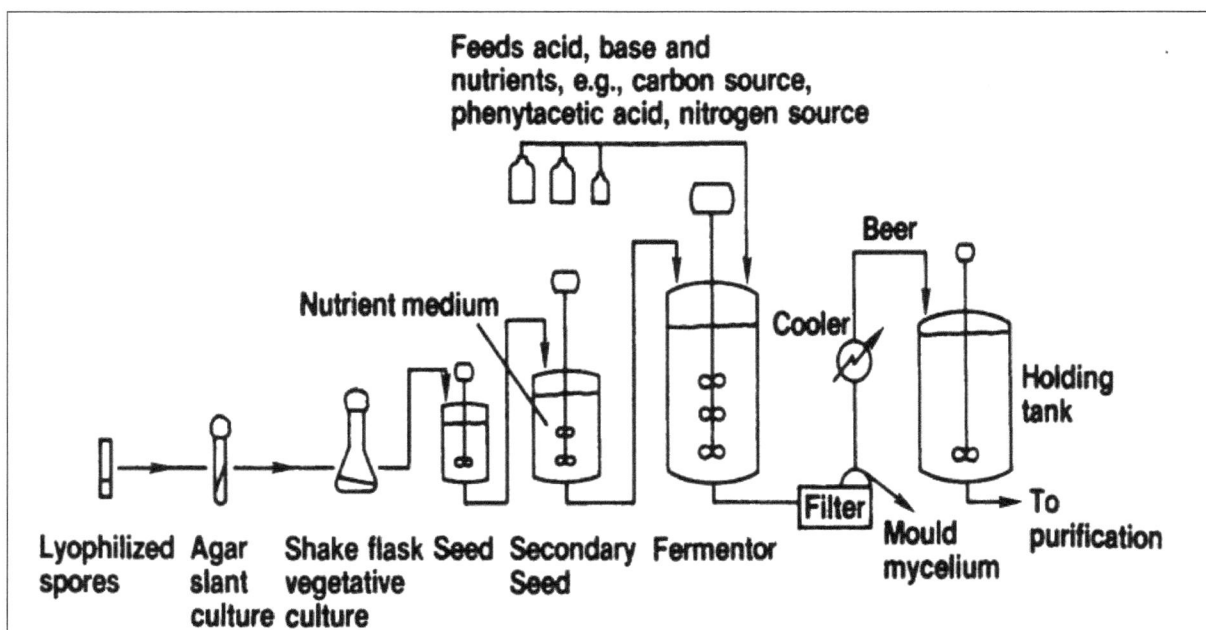

Steps involved in the production of penicillin.

Since the development of a new drug is a costly proposition, pharmaceutical companies have done very little research in the last decade. However, an alarming development has spurred a revived interest in the development of new antibiotics.

It turns out that some of the disease-causing bacteria have mutated and developed a resistance to many of the standard antibiotics. This could have grave consequences on the world's public health unless new antibiotics are discovered or improvements are made on the ones that are available. This challenging problem will be the focus of research for many years to come.

Enzymes

Micro-organisms can be used to produce enzymes that are useful in certain industrial applications. A more efficient method of using them is to isolate an enzyme and use it by itself. This is better because higher concentrations of enzyme can be obtained and only one reaction is taking place, whereas using whole micro-organisms would have numerous enzyme reactions and more products to process.

A further problem encountered is that enzymes are expensive to produce. This is solved by reusing the enzyme, this is made possible by immobilizing the enzymes, this way they do not contaminate the end products which can be reused.

Monoclonal Antibodies (mAB)

Hybridomas are cells that have been engineered to produce a desired antibody in large amounts, to produce monoclonal antibodies. Monoclonal antibodies can be produced in specialized cells through a technique now popularly known as hybridoma technology.

Hybridoma technology was discovered in 1975 by three scientists, Georges Kohler of West Germany and Cesar Milstein of Argentina and Niels Jerne of Denmark. They were awarded the 1984 Noble prize for physiology and medicine.

Methodology

A hybridoma is produced by the injection of a specific antigen into a mouse, procuring the antigen-specific plasma cells (antibody-producing cell) from the mouse's spleen and the subsequent fusion of this cell with a cancerous immune cell called a myeloma cell. The hybrid cell, which is thus produced, can be cloned to produce many identical daughter clones.

These daughter clones then secrete the immune cell product. Since these antibodies come from only one type of cell (the hybridoma cell) they are called monoclonal antibodies. The advantage of this process is that it can combine the qualities of the two different types of cells; the ability to grow continually, and to produce large amounts of pure antibody.

HAT medium (Hypoxanthine Aminopetrin Thymidine) is used for preparation of monoclonal antibodies. Laboratory animals (e.g.; mice) are first exposed to an antigen whose corresponding antibody we want to develop. Once splenocytes are isolated from the mammal, the B cells are fused with immortalized myeloma cells, which lack the HGPRT (hypoxanthine-guanine phosphoribosyltrans-ferase) gene, using polyethylene glycol or the Sendai virus.

Mehtods of Immobilizing Enzymes		
Method	How it works	Illustration
Bonding with cross- linking agent	Enzymes are linked to each other by chemical bonds. They may all be joined by amino acids for instance	
Entrapment inside a gel	Enzyme is trapped in an inert (matrix)such as alginate or collagen, and cannot be washed out. Substrates and products though, may diffuse in and out of the matrix. This type of enzyme takes the form of small beads.	
Binding to an adsorbing agent	A support matrix (glass beads or carbon) has enzymes attached to it by weak force(a bit like magnetism). The and the force might be too weak to hold the enzymes in place very strongly	

Fused cells are incubated in the HAT (Hypoxanthine-Aminopetrin Thymidine) medium. Aminopetrin in the myeloma cells die, as they cannot produce nucleotides by the de novo, or salvage medium blocks the pathway that allows for nucleotide synthesis.

Hence, un-fused D cells die. Un-fused B cells die as they have a short life span. Only the B cell-myeloma hybrids survive, since the HGPRT gene coming from the B cells is functional. These cells produce antibodies (a property of B cells) and are immortal (a property of myeloma cells).

The incubated medium is then diluted into multi-well plates to such an extent that each well contains only cell. Then the supernatant in each well can be checked for desired anti body. Since the antibodies in a well are produced by the same B cell, they will be directed towards the same epitope, and are known as monoclonal antibodies.

Once a hybridoma colony is established, it will continually grow in culture medium like RPMI-1640 (with antibiotics and foetal bovine serum) and produce antibody.

The next stage is a rapid primary screening process, which identifies and selects only those hybridomas that produce antibodies of appropriate specificity. The hybridoma culture supernatant, secondary enzyme labelled conjugate, and chromogenic substrate, is then incubated, and the formation of a coloured product indicates a positive hybridoma.

Alternatively, immunocytochemical screening can also be used. Multi-well plates are used initially to grow the hybridomas and after selection, are changed to larger tissue culture flasks. This maintains the well-being of the hybridomas and provides enough cells for cryopreservation and supernatant for subsequent investigations.

The culture supernatant can yield 1 to 60 ug/ml of monoclonal antibody, which is maintained at 20 °C or lower until required. By using culture supernatant or a purified immunoglobulin preparation, further analysis of a potential monoclonal antibody producing hybridoma can be made in terms of reactivity, specificity, and cross-reactivity.

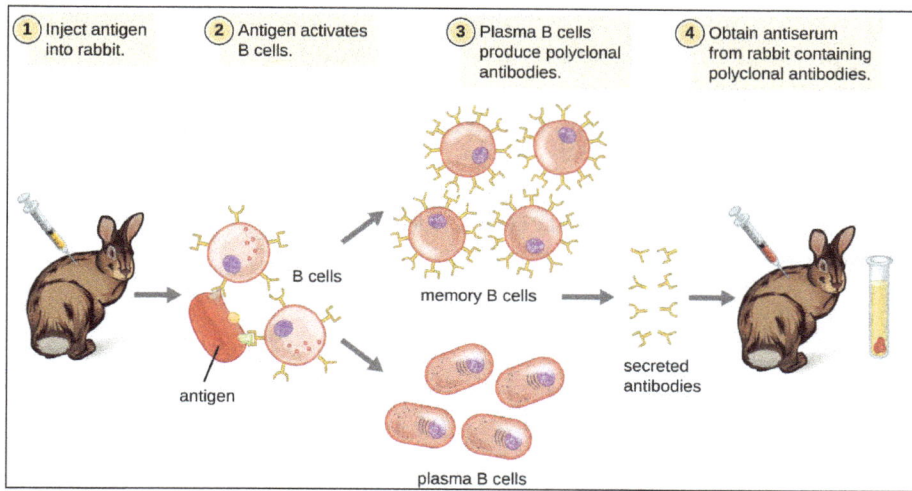

Production of monoclonal antibodies.

Large-Scale Production of Monoclonal Antibodies (mAB)

Large scale production of monoclonal antibodies consider production scales of 0.1-10 g as small, 10-100 gas medium, and over 100 g as large. This is generally performed to produce monoclonal antibodies for three purposes: diagnosis, therapy, and research on and development of new therapeutic agents.

Amino Acid Production

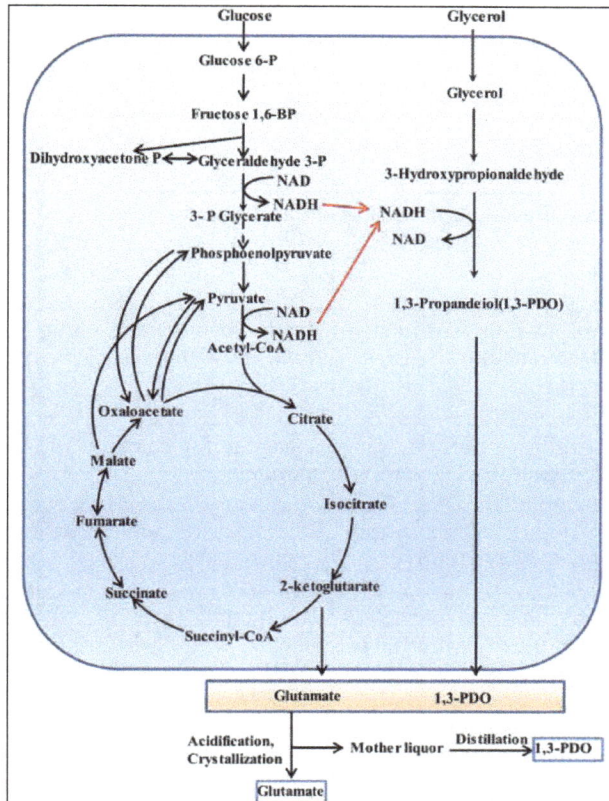

Production of Glutamate from the mutant species of Corynebacterium glutamicum.

Amino acids such as lysine and glutamic acid have been used in the food industries for a long time as a nutritional supplement (in order to increase the nutritional value of the bread) and as a flavor enhancing compound. Glutamic acid is produced in large quantities by using mutants of Corynebacteriumglutamicum that lack the ability to convert the TCA cycle intermediate alpha ketoglutarate to succinyl-CoA.

Lysine is also produced from the mutant bacterium Corynebacteriumglutamicum which blocks the synthesis of homoserine and accumulates lysine.

Organic Acid Production

Organic acids such as Citric, acetic, lactic, fumaric, and gluconic acids are major products of microbial industry. Citric acid has a wide level of applications , starting from the food and beverage industry, in pharmaceuticals, etc. Aspergillusniger is used for the production of citric acid.

The fundamental idea behind the production of citric acid is to limit the availability of the amounts of trace metals such as manganese and iron to prevent the growth of Aspergillusniger at a specific point in the process of fermentation.

The medium for the citric acid fermentation contains high sugar concentrations (15 to 18%) and copper.The success of this fermentation depends widely on the regulation and functioning of the glycolytic pathway and the tricarboxylic acid cycle.

Major Organic Acids Produced buy Microbial Processes			
Product	Microorganism used	Eepresentative uses	Fermentation conditions
Acetic acid	Acetobacter with ethanol solutions	Wide variety of foods uses	Single- step oxidation, with 15% solutions produced; 95.99%yields
Citric acid	Aspergillus niger in molasses-based medium	Pharmaceuticals, as a food additive	High carbohydrate concentrations and controlled limitation of trace metals; 60.80% yields
Funmarice acid	Rhizopus nigricans in sugar based medium	Resin manufacture, tanning, and sizing	Strongly aerobic fermentation;carbon-nitrogen ratio is critical; zinc should be limited; 60% yields
Gluconic acid	Aspergillus niger in glucose-mineral salts medium	A carrier for calcium and sodium	Uses agitation or stirred fermenters; 95% yields
Itaconic acid	Aspergillus terreus in molasses-salts medium	Esters can be polymerized to make plastics	Highly aerobic medium, below pH 2.2;85% yields
Kojic acid	Aspergillus flavus-oryzae in carbohydrate-inorganic N medium	The manufacture of fungicides and insecticides when comple-xed with metals	Iron must be carefully controlled to avoid reaction with kojic acid alter fermentation
Lactic acid	Homofementative Lactobacillus delbrueckii	As a crrier for calcium and as an acidifier	Purified medium used to facilitate extraction

Cheese Production

Cheese is produced from a lactic acid fermentation of milk, which results in coagulation of milk proteins and formation of a curd. Rennin is generally used to promote this curd formation. After the curd is formed, it is heated and pressed in order to remove the watery part of the milk called as whey.

The solid curd is then salted and kept aside for ripening. In the production of cheese, microorganisms are widely used. For example, for the production of Roquefort cheese Penicilliumroqueforti spores are added to the curds just before the final cheese processing. Also during the production of Camembert cheese, the cheese is inoculated with spores of Penicilliumcamemberti.

Production of Alcoholic Beverages

Alcoholic beverages are basically produced from the fermentation of carbohydrates. The most widely consumed alcoholic beverages are wine and beer.

Production of Wine

Wine is produced by the fermentation of grape juice. Folio wings are the basic six steps involved in the production of wine.

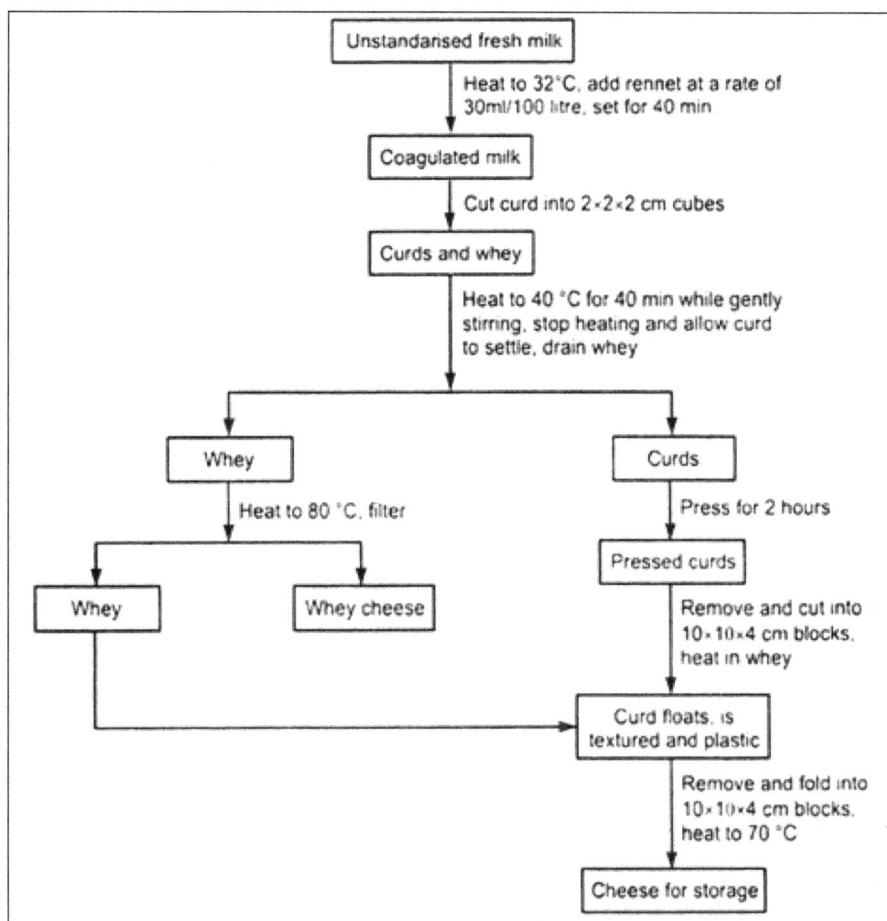

Step involved in the production of cheese.

Step 1: Viticulture

As a known fact the flavour of the wine is dependent on the kind of grapes which are used. The variety of grapes depends on the place where they are grown, the climate of the place where they are grown, the drainage system around the winery, the quality of the soil, humidity content of the region and even the exposure levels to the sun.

Other than these factors, one of the most vital points which is an important determinant in the making of wine is the techniques used to make the wine. In fact, the different wineries formulate a customized wine making procedure according to the needs of their winery, so that they can bring out the best flavours of their grapes.

Step 2: Harvesting

The next step in the wine-making process is the harvesting of the finely cultivated grapes. When the grapes are ripped and harvested it is very important that the timing of ripping the grapes is apt and the exact timing can be ascertained through experience.

The timing of the ripping the grapes should be such that the grapes have the apt combination of sugar, acid and moisture. The harvesting of the grapes can be either done manually or mechanically, however majority of the grape wineries reap it manually.

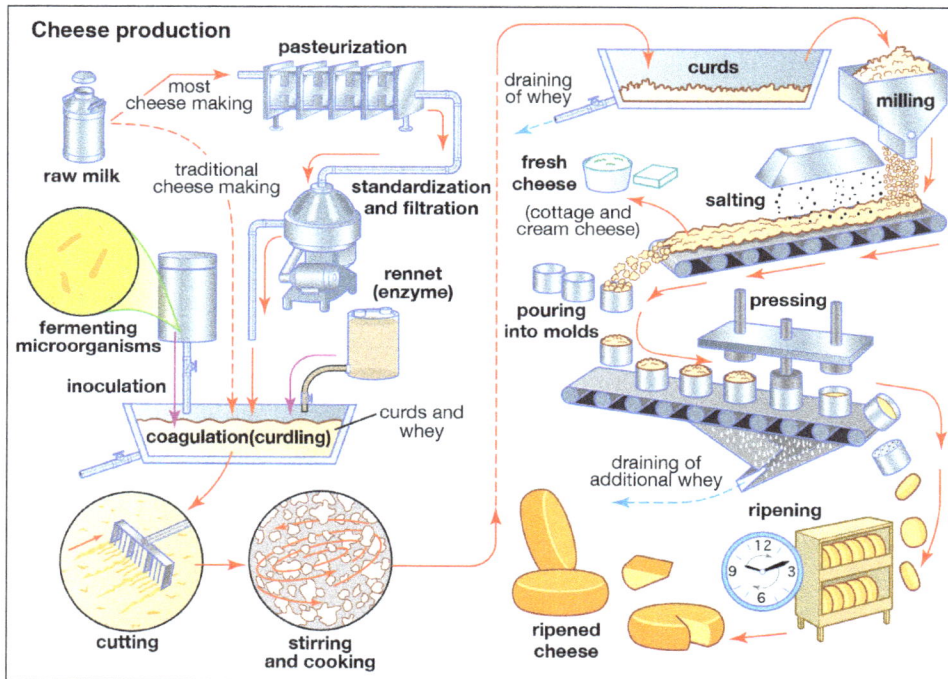

Production of cheese

Step 3: Crushing

The next step after the grapes have been harvested is to crush and press so as to have their inherent flavours in the form of liquid. As the fruits Eire crushed the grapes withdraw their moisture content and the sugars.

Many wineries make the usage of specialized machines for the process of crushing and pressing the grapes. The resultant liquid which is formed after crushing and pressing the wines is referred to as must, at this stage the wine will either turn into red colour or white colour depending on the preference of the wine-maker.

The wines become red when they are simply left after, they are crushed because in this stage juices from the skin and the flavours are ripped off, thus when you leave the must for a certain period of time, the wine become red wine, and in case you immediately crush the wine after pressing and segregate the skin, then the wine becomes white wine.

Step 4: Fermentation

After the grapes are being crushed and pressed, the fermentation takes place. Since the grapes have good quantity of sugar and moisture they easily get fermented with the reaction of wild yeast. Fermentation takes place in almost 10 to 30 days; however, this depends on the quality of grapes and the climate.

Step 5: Clarifying the Solution

The next step in wine making is clarifying the solution. It is also referred to as stabilization.

Step 6: Bottling

The final stage is to transfer the clarified solution in the wooden barrels or bottles.

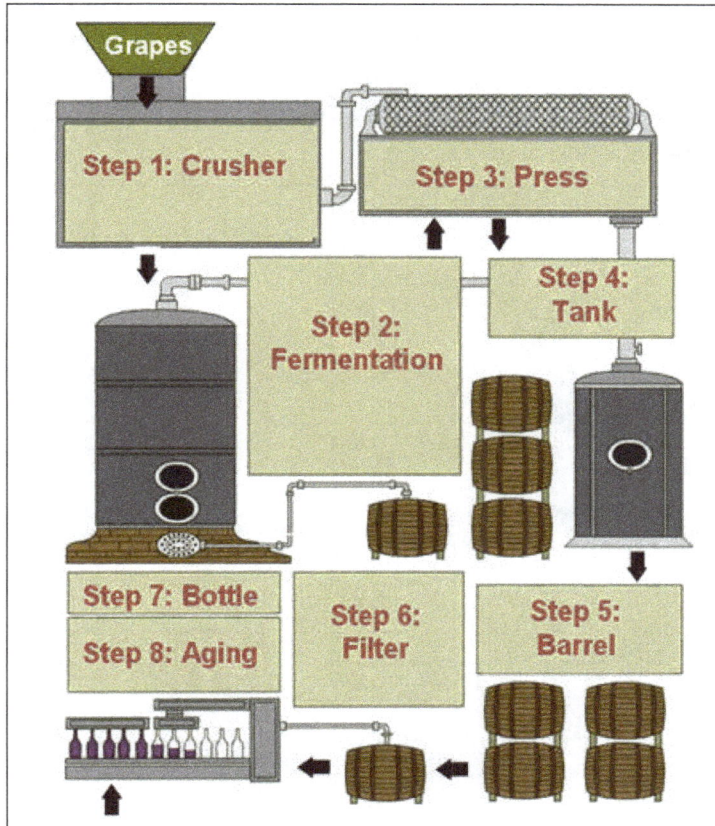

Steps involved in the production of wine.

Production of Beer

Step 1: Extraction

Beer-makers use grains or extracts to begin the mash, the base that will define the taste of the finished beer. Grains of barley or wheat are pre- soaked to begin the release of enzymes to produce maltose, a sugar. The brewer crushes a mixture of malted and un-malted grains and then heats them in water.

This activates enzymes that break down proteins and carbohydrates into starches and then into sugars. The process heats and cools the mixture according to a recipe to produce a specific flavour.

When finished, the liquid is drained through the sediment and the grains are dried in a kiln. The dried grain is soaked in warm water and drained again to extract more sugars from the mash. A final rinse with fresh hot water pulls out the last of the sugars.

The three extractions produce a liquid called "wort," which is brought to the boil to kill any harmful bacteria and concentrate the extract. Hops, a cone-like bloom, contains oils and resins that offset the sweetness of the sugars. It may be added to the wort during the "boil" to "bitterize" the mash before fermentation or it may be added at the end of fermentation, depending on the recipe and brewer.

Each mix of grain and hops produce different types and amounts of sugars to be available for fermentation, the next process in brewing.

Step 2: Fermentation

Fermentation is a chemical process. The brewer adds yeast (Saccharomyces cerevisiae), a type of fungus that metabolizes sugars in the absence of oxygen, to the aerated wort. As fermentation begins, a foamy coat of "krausen" grows on the surface of the wort and the chemical reaction produces carbon dioxide and alcohol.

Steps involved on the production of Beer.

A valve on top of the sealed fermentation chamber vents the carbon dioxide as the fermentation progresses.

The chemical reaction also produces esters and phenols that contribute to the character of the beer's flavour as well as chlorine and fatty acids. Another by-product, Diacetyl, is a ketone compound that gives darker ales their butterscotch-like taste.

Lactic acid is often added to wheat wort to add tartness. Temperature is used to control the fermentation process, but eventually it will stop because the sugars have been consumed and most of the yeasts have died.

Step 3: The Finish

The final process of brewing is bottling but many brewers insert intermediate steps in this finishing process. The fermented wort must be separated from the foamy krausen and solid remnants of hops, dead yeast and other by-products that may coat the sides of the chamber, often by draining from the bottom or siphoning.

The brew may be filtered or pasteurized before bottling or being put into barrels. As beer is bottled or barrelled, a bit of sugar—white sucrose, brown or molasses—is added to it to encourage the remaining live yeast to ferment and vent carbon dioxide, which will stay in the sealed container as carbonation.

Permissions

All chapters in this book are published with permission under the Creative Commons Attribution Share Alike License or equivalent. Every chapter published in this book has been scrutinized by our experts. Their significance has been extensively debated. The topics covered herein carry significant information for a comprehensive understanding. They may even be implemented as practical applications or may be referred to as a beginning point for further studies.

We would like to thank the editorial team for lending their expertise to make the book truly unique. They have played a crucial role in the development of this book. Without their invaluable contributions this book wouldn't have been possible. They have made vital efforts to compile up to date information on the varied aspects of this subject to make this book a valuable addition to the collection of many professionals and students.

This book was conceptualized with the vision of imparting up-to-date and integrated information in this field. To ensure the same, a matchless editorial board was set up. Every individual on the board went through rigorous rounds of assessment to prove their worth. After which they invested a large part of their time researching and compiling the most relevant data for our readers.

The editorial board has been involved in producing this book since its inception. They have spent rigorous hours researching and exploring the diverse topics which have resulted in the successful publishing of this book. They have passed on their knowledge of decades through this book. To expedite this challenging task, the publisher supported the team at every step. A small team of assistant editors was also appointed to further simplify the editing procedure and attain best results for the readers.

Apart from the editorial board, the designing team has also invested a significant amount of their time in understanding the subject and creating the most relevant covers. They scrutinized every image to scout for the most suitable representation of the subject and create an appropriate cover for the book.

The publishing team has been an ardent support to the editorial, designing and production team. Their endless efforts to recruit the best for this project, has resulted in the accomplishment of this book. They are a veteran in the field of academics and their pool of knowledge is as vast as their experience in printing. Their expertise and guidance has proved useful at every step. Their uncompromising quality standards have made this book an exceptional effort. Their encouragement from time to time has been an inspiration for everyone.

The publisher and the editorial board hope that this book will prove to be a valuable piece of knowledge for students, practitioners and scholars across the globe.

Index

www.ingramcontent.com/pod-product-compliance
Lightning Source LLC
Chambersburg PA
CBHW061302190326
41458CB00011B/3740

* 9 7 8 1 6 4 1 7 2 6 1 2 2 *